Rivals on the Road

The Ford  Chevrolet

Etienne Psaila

Rivals on the Road: The Ford vs. Chevrolet Story

Copyright © 2024 by Etienne Psaila. All rights reserved.

First Edition: **November 2024**

No part of this publication may be reproduced, distributed, or transmitted in any form or by any means, including photocopying, recording, or other electronic or mechanical methods, without the prior written permission of the publisher, except in the case of brief quotations embodied in critical reviews and certain other non-commercial uses permitted by copyright law.

ISBN: 978-1-7638074-6-4

Table of Contents

Introduction: The Birth of an American Rivalry
Chapter 1: Pioneers of the Open Road
Chapter 2: The Roaring Twenties to the Great Depression
Chapter 3: Post-War Prosperity and the Rise of the American Dream
Chapter 4: Enter the Muscle Cars
Chapter 5: The Ford Mustang—A New Breed of Pony Car
Chapter 6: Chevrolet Strikes Back—The Birth of the Camaro
Chapter 7: The Pony Car Wars—Mustang vs. Camaro
Chapter 8: Battlegrounds Beyond the Showroom—Motorsports
Chapter 9: Trucks and Workhorses—F-Series vs. Silverado
Chapter 10: Navigating the Oil Crisis and Environmental Concerns
Chapter 11: The Evolution of Design and Technology in the '80s and '90s
Chapter 12: The Revival of the Classics
Chapter 13: Globalization and International Markets
Chapter 14: The Digital Age—Marketing and Community Building
Chapter 15: Innovations in Sustainability and Technology
Chapter 16: The Impact on Pop Culture
Chapter 17: The Enthusiasts and the Aftermarket Scene
Chapter 18: Case Studies—Notable Models and Their Stories
Chapter 19: Challenges and Controversies
Chapter 20: Looking Ahead—The Future of the Rivalry
Conclusion: A Legacy Driven by Competition

Introduction: The Birth of an American Rivalry

The hum of machinery filled the air as a young Henry Ford tinkered late into the night in a small brick shed behind his Detroit home. It was the late 1890s, and the world stood on the cusp of a transportation revolution. Across the Atlantic, a Swiss-born mechanic and race car driver named Louis Chevrolet was honing his skills, unaware that his path would soon cross with the burgeoning American automobile industry. These two men, from vastly different backgrounds, were about to ignite a rivalry that would shape not only an industry but the very fabric of American culture.

The Visionary from Dearborn

Henry Ford was born on a farm in Dearborn, Michigan, in 1863. From an early age, he exhibited a fascination with mechanics, often disassembling and reassembling watches to understand their inner workings. This curiosity propelled him into a career that began at the Edison Illuminating Company, where he worked as an engineer. However, Ford's true passion lay in creating a horseless carriage that could be accessible to the masses.

In 1903, after a couple of unsuccessful ventures, Ford founded the Ford Motor Company. His mission was clear: to produce an affordable automobile for the average American. This vision materialized with the introduction of the Model T in 1908. Priced at $825, the Model T was revolutionary—not just for its engineering but for what it represented. It was the embodiment of mobility and freedom, symbols that resonated deeply with the American spirit.

Ford didn't just stop at creating an affordable car; he revolutionized the manufacturing process itself. In 1913, he introduced the moving assembly line, drastically reducing production time and costs. This innovation made the Model T even more affordable, dropping the price to as low as $260 by the 1920s. Ford's approach democratized automobile ownership, turning cars from luxury items into necessities.

The Racer Turned Automaker

Meanwhile, Louis Chevrolet was carving out his own path. Born in La Chaux-de-Fonds, Switzerland, in 1878, Chevrolet was the son of a watchmaker. His mechanical aptitude led him to France and later to Canada, where he worked as a mechanic and chauffeur. His skills behind the wheel caught the attention of racing enthusiasts, and soon he was competing in races across America.

In 1911, Chevrolet teamed up with William C. Durant, the ousted founder of General Motors, to establish the Chevrolet Motor Car Company. Durant saw in Chevrolet a marketable name and a skilled engineer who could help him re-enter the automotive industry. Their first car, the Classic Six, was a robust vehicle that catered to the upper end of the market, a direct contrast to Ford's utilitarian Model T.

However, Durant recognized the need to compete with Ford's affordability. By 1915, Chevrolet shifted focus to produce more economical cars, leading to the creation of the Model 490, named after its $490 price tag, intentionally set to undercut the Model T. This strategic move positioned Chevrolet as a direct competitor to Ford, setting the stage for a rivalry that would intensify over the coming decades.

Sowing the Seeds of Competition

The early 20th century was a period of rapid industrial growth and innovation in America. The automobile industry was at the forefront of this transformation, symbolizing the nation's ingenuity and entrepreneurial spirit. Ford and Chevrolet, through their respective companies, became emblematic of these ideals.

Ford's emphasis on affordability and mass production made automobiles accessible, but it also meant limited options for consumers seeking variety or luxury. Chevrolet capitalized on this gap by offering vehicles that combined affordability with added features and style. This approach appealed to consumers who desired more than just basic transportation.

The competition extended beyond pricing and features; it was also about brand identity and loyalty. Ford's Model T was affectionately known as the "Tin Lizzie," a reliable workhorse that became a fixture in American households. Chevrolet, on the other hand, cultivated an image of performance and sophistication, partly influenced by Louis Chevrolet's racing background.

Shaping American Culture and Identity

The rivalry between Ford and Chevrolet transcended the automotive industry; it became a cultural phenomenon. It encapsulated the essence of the American Dream—the belief that innovation, hard work, and competition could lead to success and prosperity.

This competition spurred technological advancements and improved quality, benefiting consumers and fueling the growth of the American economy. The companies' efforts to outdo each other led to better products and services, embodying the capitalist ideals upon which the nation was built.

Moreover, the rivalry influenced social dynamics. Ownership of a Ford or a Chevrolet became a statement of personal identity, reflecting individual values and preferences. Communities formed around brand loyalty, fostering a sense of belonging and camaraderie among owners.

The Ford vs. Chevrolet debate permeated popular culture,

appearing in music, films, and literature. It became a topic of discussion around dinner tables and in mechanic shops, symbolizing deeper themes of tradition versus innovation, reliability versus style, and mass production versus customization.

A Legacy Begins

As the first half of the 20th century unfolded, the foundations laid by Henry Ford and Louis Chevrolet evolved into a full-fledged rivalry. Their companies became titans of industry, each pushing the other to new heights. This competition not only drove the advancement of automotive technology but also mirrored the dynamic and competitive nature of American society.

The story of Ford and Chevrolet is more than a tale of two car manufacturers; it's a reflection of America's journey through innovation, adversity, and triumph. It's about how competition can inspire greatness and how rivalries can foster progress that benefits all.

As we delve deeper into the chapters ahead, we'll explore how this rivalry unfolded across different eras, influenced consumer choices, and left an indelible mark on motorsports and popular culture. From the roaring engines of muscle cars to the cutting-edge technologies of today, the Ford vs. Chevrolet story is a testament to the power of competition in driving excellence.

Chapter 1: Pioneers of the Open Road

The dawn of the 20th century was a time of profound transformation. Cities buzzed with newfound energy, industries burgeoned, and the promise of modernity beckoned from every corner. Amidst this backdrop of change, the automobile emerged as a symbol of progress and freedom. It was in this fertile ground that Ford and Chevrolet began their ascent, forever altering the landscape of transportation and society itself.

The Rise of the Model T: Ford's Masterstroke

In 1908, the Ford Motor Company unveiled the Model T, a car that would not only define the company but also revolutionize the automotive industry. Priced initially at $825, the Model T was designed with a singular vision: to be affordable, durable, and easy to maintain for the average American.

Henry Ford believed that mobility should not be a luxury reserved for the wealthy. He famously stated, "I will build a motor car for the great multitude." To achieve this, Ford focused on simplicity in design and efficiency in production. The Model T featured a robust four-cylinder engine, a planetary transmission, and a high ground clearance suitable for the unpaved roads of rural America.

The real innovation, however, lay not just in the car itself but in how it was produced. In 1913, Ford introduced the moving assembly line at his Highland Park plant. This groundbreaking method reduced the assembly time of a single car from over 12 hours to just 93 minutes. Workers remained stationary as the vehicle moved along the line, with each person performing a specific task repeatedly. This not only increased productivity but also lowered costs significantly.

By passing the savings onto consumers, the price of the

Model T steadily decreased. By 1925, it cost a mere $260, making it accessible to millions. The Model T became affectionately known as the "Tin Lizzie," and it wasn't just a car—it was a phenomenon. By the end of its production in 1927, over 15 million units had been sold, cementing its place in history.

Chevrolet Enters the Fray

While Ford was dominating the market with the Model T, a new competitor was gearing up to challenge its supremacy. The Chevrolet Motor Car Company, founded in 1911 by Louis Chevrolet and William C. Durant, sought to create vehicles that combined performance with affordability.

Louis Chevrolet brought his expertise as a mechanic and racer to the company, influencing the design of their early models. The first Chevrolet car, the Series C Classic Six, was introduced in 1912. It was a well-appointed vehicle with a powerful six-cylinder engine, electric starter, and electric headlights—a luxury at the time. However, its $2,150 price tag placed it out of reach for most consumers.

Recognizing the need to compete with Ford's pricing, Durant shifted the company's strategy. In 1915, Chevrolet introduced the Model 490, named for its $490 price point. The Model 490 was a direct competitor to the Model T, offering additional features like electric lights and a more refined interior. This aggressive pricing and added value began to attract consumers who desired more than just basic transportation.

In 1918, Chevrolet became part of General Motors (GM), with Durant returning to lead GM after a brief ousting. Under GM, Chevrolet had access to greater resources and technology, allowing it to expand its lineup and production capabilities.

Innovation Drives Competition

The intense competition between Ford and Chevrolet spurred significant advancements in automobile technology and manufacturing. While Ford focused on perfecting the assembly line and reducing costs, Chevrolet emphasized offering more features and comfort at competitive prices.

Ford's insistence on the Model T's simplicity eventually became a liability. By the 1920s, consumer preferences began shifting towards cars that offered style, comfort, and customization. Chevrolet capitalized on this by introducing models with closed bodies, multiple color options, and improved interiors, all while keeping prices within reach of the average buyer.

In response, Ford attempted to modernize the Model T but was reluctant to deviate from his philosophy of simplicity and uniformity. It wasn't until 1927 that Ford discontinued the Model T and introduced the Model A, acknowledging the need for innovation to meet consumer demands.

Democratizing Car Ownership

The early innovations of Ford and Chevrolet did more than create successful companies; they democratized car ownership in America. The accessibility of the Model T and the competitive offerings from Chevrolet meant that automobiles were no longer exclusive to the affluent. Families across the nation could now enjoy the freedom and convenience of personal transportation.

This widespread car ownership had profound societal impacts. It facilitated suburban expansion, as people were no longer confined to living near workplaces accessible by foot or public transit. Rural communities became less isolated, with farmers and small-town residents gaining easier access to markets, goods, and services.

The automobile also transformed the economy. Industries related to car production and maintenance—steel, rubber, glass, oil, road construction—experienced significant growth. Job creation in these sectors contributed to the overall prosperity of the nation.

Setting Industry Standards

The strategies and innovations introduced by Ford and Chevrolet set new standards for the automotive industry. Ford's assembly line became a model for mass production worldwide, influencing manufacturing processes in various industries. The focus on efficiency and cost reduction without sacrificing quality became a guiding principle for manufacturers.

Chevrolet's approach of offering more features and customization at affordable prices established the importance of meeting consumer desires beyond basic functionality. This strategy highlighted the significance of market research and understanding consumer trends, which became essential practices in product development.

Both companies also emphasized the importance of dealer networks and customer service. Ford and Chevrolet dealerships became ubiquitous across the country, providing not just sales but also maintenance and repair services. This accessibility further encouraged car ownership by ensuring that customers had support after their purchase.

The Open Road Beckons

As automobiles became integral to American life, the open road transformed from a mere concept into a tangible reality. The newfound mobility allowed people to explore beyond their immediate surroundings, giving rise to road trips, tourism, and a sense of adventure.

The demand for better roads led to significant infrastructure developments. The Federal Aid Road Act of 1916 and subsequent legislation provided funding for highway construction, recognizing the critical role of transportation in economic growth and national cohesion.

Automobiles also began to influence culture and societal norms. They provided young people with unprecedented independence, affected dating practices, and even impacted architectural designs with the advent of drive-in restaurants and theaters.

Challenges and Adaptations

The rapid growth of the automobile industry was not without challenges. Issues such as traffic congestion, accidents, and environmental concerns began to emerge. Both Ford and Chevrolet had to navigate these challenges while continuing to innovate and meet consumer demands.

Labor relations became a significant issue, particularly for Ford. The monotonous work on assembly lines led to high employee turnover. In response, Ford introduced the $5 workday in 1914, doubling the average wage and setting new labor standards. This move improved worker satisfaction and loyalty but also drew criticism from competitors and some sectors of society.

Chevrolet, under GM's umbrella, benefited from Alfred P. Sloan's leadership. Sloan introduced the concept of planned obsolescence and annual model changes, encouraging consumers to purchase new cars more frequently. He also implemented decentralized management, allowing different divisions within GM to operate semi-autonomously, fostering innovation and responsiveness to market changes.

Legacy of the Early Years

The pioneering efforts of Ford and Chevrolet in the early 20th century laid the foundation for the modern automotive industry. Their approaches to manufacturing, marketing, and customer engagement became benchmarks for others to follow.

The competition between the two companies fueled continuous improvement, ensuring that the benefits of innovation reached consumers. Their commitment to making cars accessible transformed not only how people traveled but also how they lived, worked, and interacted.

As the nation moved forward, the automobiles produced by Ford and Chevrolet became more than just vehicles; they became symbols of progress, freedom, and the American way of life. The open road was no longer a distant dream but an invitation to explore, connect, and embrace the possibilities of a rapidly changing world.

Chapter 2: The Roaring Twenties to the Great Depression

The 1920s, often referred to as the Roaring Twenties, was a decade marked by economic prosperity, cultural flourishing, and significant technological advancements in the United States. The automobile industry, spearheaded by giants like Ford and Chevrolet, played a pivotal role in this era of exuberance. However, the euphoria was short-lived. The onset of the Great Depression in 1929 plunged the nation into economic despair, challenging industries to adapt and survive. This chapter explores how Ford and Chevrolet navigated these turbulent times, their competitive strategies, and how they endeavored to maintain sales and customer loyalty amid unprecedented economic downturns.

The Roaring Twenties: A Decade of Prosperity and Innovation

Ford's Dominance and Stagnation

As the 1920s dawned, Ford Motor Company was at the zenith of its power. The Model T had become a ubiquitous presence on American roads, with production numbers soaring. By 1921, Ford held a staggering 60% share of the automobile market. The company's success was rooted in Henry Ford's mastery of mass production and his commitment to affordability.

However, the very factors that propelled Ford to dominance began to sow seeds of stagnation. Henry Ford was deeply committed to the Model T, resisting changes to its design and features. He believed that the car was the epitome of what the public needed—a reliable and inexpensive means of transportation. This reluctance to innovate led to a product line that lacked variety and modernization.

Chevrolet's Strategic Innovations

In contrast, Chevrolet, under the General Motors (GM) umbrella and the leadership of Alfred P. Sloan Jr., adopted a different approach. Recognizing the limitations of Ford's one-model strategy, Chevrolet introduced annual model changes, offering updated designs, features, and options that catered to evolving consumer tastes.

Sloan implemented the concept of "a car for every purse and purpose," positioning Chevrolet as the entry-level brand within GM's hierarchy. This strategy provided consumers with a progression path as their incomes and preferences changed. Chevrolet's vehicles were marketed not just as transportation but as symbols of status and personal expression.

Technological and Design Advancements

Chevrolet invested heavily in technological innovations and aesthetic improvements. Features like electric starters, closed bodies, and improved braking systems were incorporated into their models. The company also offered a wider range of colors and styles, appealing to consumers seeking individuality.

In 1924, Chevrolet introduced the "Superior" series, which boasted advancements in engine performance and comfort. These cars were competitively priced against the Model T but offered more in terms of amenities and style. Chevrolet's focus on customer desires began to erode Ford's market share.

Marketing and Consumer Appeal

Chevrolet leveraged aggressive marketing campaigns to highlight the advantages of their vehicles over Ford's. Advertisements emphasized quality, style, and modernity, appealing to the aspirations of the burgeoning middle class.

One notable campaign featured the slogan, "A Six for the Price of a Four," promoting their six-cylinder engines as superior to Ford's four-cylinder models. This messaging resonated with consumers who were becoming more discerning and desired vehicles that reflected their success and sophistication.

Ford's Response and the Introduction of the Model A

Realizing the need to adapt, Henry Ford reluctantly agreed to discontinue the Model T in 1927 after 19 years of production. The company shut down its plants for six months to retool for the new Model A. This hiatus allowed competitors like Chevrolet to gain ground, but Ford believed the overhaul was necessary.

The Model A: A New Direction

The Model A represented a significant departure from Ford's previous philosophy. It featured modern amenities such as safety glass, hydraulic shock absorbers, and a more powerful engine. The car was available in various body styles and colors, addressing consumer demand for variety.

The Model A was well-received, with nearly 5 million units sold by the end of its production in 1931. However, the delay in responding to market changes had allowed Chevrolet to strengthen its position, and the competitive landscape had shifted.

The Great Depression: Navigating Economic Turbulence

The Market Collapse

The stock market crash of 1929 ushered in the Great Depression, a period of severe economic contraction.

Unemployment soared, consumer spending plummeted, and industries across the board faced declining sales.

The automobile industry was particularly hard-hit. Cars were a significant investment, and with incomes dwindling, consumers postponed purchases. Both Ford and Chevrolet faced the daunting task of sustaining operations and retaining customer loyalty in an environment where demand had evaporated.

Ford's Challenges and Strategies

Ford Motor Company responded to the downturn by cutting production and reducing prices. However, Henry Ford's aversion to external financing and his insistence on self-reliance limited the company's flexibility. He was reluctant to embrace new technologies or marketing strategies that required additional capital investment.

Labor relations at Ford deteriorated during this period. The company's workforce faced wage cuts and layoffs, leading to unrest and strikes. Henry Ford's anti-union stance exacerbated tensions, culminating in violent confrontations such as the "Battle of the Overpass" in 1937.

Despite these challenges, Ford continued to invest in innovations. In 1932, the company introduced the Flathead V8 engine, making it the first affordable V8-powered car on the market. This advancement provided superior performance and was a significant selling point, helping to attract customers seeking value in tough economic times.

Chevrolet's Adaptability and Consumer Focus

Chevrolet, under GM's leadership, implemented a more flexible approach. Recognizing the need to stimulate demand, GM introduced the "Installment Plan," allowing customers to purchase vehicles on credit with manageable monthly payments. This financing option made car ownership attainable even during economic hardship.

Chevrolet also continued its emphasis on annual model updates and technological improvements. In 1933, the company introduced the "Knee-Action" independent front suspension, enhancing ride comfort and handling. These innovations were marketed aggressively, highlighting Chevrolet's commitment to providing modern, value-packed vehicles.

Moreover, GM's diversified portfolio allowed it to allocate resources strategically across its brands. This financial stability enabled Chevrolet to maintain production levels, invest in advertising, and offer incentives to dealers and customers.

Marketing and Brand Loyalty

Both companies recognized the importance of maintaining brand visibility and customer engagement during the Depression. Chevrolet's advertising campaigns focused on themes of reliability, economy, and innovation. Slogans like "The Great American Value" emphasized the company's understanding of consumers' financial constraints.

Ford, while less aggressive in marketing, relied on the reputation of its brand and the appeal of the V8 engine. The company produced promotional films and sponsored radio programs to keep the Ford name in the public consciousness.

Dealer Networks and Customer Service

The strength of dealer networks became crucial during this period. Chevrolet encouraged its dealers to offer exceptional customer service, flexible financing options, and community engagement activities. Dealerships hosted events, provided maintenance services, and built relationships that fostered loyalty.

Ford dealers, facing directives from a more centralized and less flexible corporate structure, struggled to match these initiatives. The disparity in dealer support contributed to Chevrolet gaining a competitive edge in customer satisfaction.

Emergence from the Depression

By the late 1930s, signs of economic recovery began to appear. Government programs under the New Deal stimulated infrastructure projects and job creation, slowly revitalizing consumer confidence.

Ford's Late Adaptations

Acknowledging the shifting dynamics, Ford began to modernize its operations and management practices. In 1937, the company introduced the "Ford Deluxe," featuring updated styling and amenities to compete with Chevrolet's offerings.

Henry Ford's son, Edsel Ford, played a significant role in pushing for innovation and modernization within the company. His efforts led to improved designs and the incorporation of customer feedback into product development.

Chevrolet's Continued Innovation

Chevrolet continued to innovate, introducing new models with advanced features. The 1939 Chevrolet Master Deluxe offered improved safety features, streamlined designs, and enhanced performance. The company also expanded its marketing efforts, sponsoring popular radio shows and participating in community initiatives.

Impact on Market Share

By the end of the 1930s, Chevrolet had overtaken Ford as the industry leader in sales. The combination of adaptable strategies, customer-focused innovations, and effective marketing had paid off. Chevrolet's ability to read the market and respond swiftly positioned it favorably as the nation headed into a new decade.

Lessons Learned and Lasting Impacts

The experiences of Ford and Chevrolet during the 1920s and 1930s underscore the importance of adaptability, innovation, and customer engagement in the face of economic fluctuations.

Adaptability and Innovation

Chevrolet's willingness to embrace change, invest in new technologies, and update models annually kept the brand fresh and appealing. This adaptability allowed the company to meet consumer demands and adjust to economic realities.

Ford's initial reluctance to move beyond the Model T demonstrated the risks of stagnation. However, the eventual introduction of the Model A and the V8 engine showcased the company's capacity for innovation when compelled to act.

Customer-Centric Strategies

Understanding and responding to consumer preferences proved critical. Chevrolet's focus on offering value-added features and financing options aligned with customer needs during the Depression. Ford's emphasis on mechanical excellence with the V8 engine appealed to performance-conscious buyers but lacked the broader customer engagement strategies employed by Chevrolet.

Marketing and Brand Loyalty

Effective marketing and fostering brand loyalty were vital in maintaining sales during economic downturns. Chevrolet's aggressive advertising and dealer support built strong customer relationships. Ford's reliance on brand reputation without equivalent marketing efforts resulted in lost ground to competitors.

Conclusion: A Decade of Transformation

The period from the Roaring Twenties to the Great Depression was one of significant transformation for both Ford and Chevrolet. The economic boom provided opportunities for expansion and innovation, while the Depression tested the resilience and adaptability of both companies.

Chevrolet emerged from the 1930s with a strengthened market position, attributed to its flexible strategies and customer-focused approach. Ford learned valuable lessons about the necessity of innovation and responsiveness to market changes.

The rivalry between the two automakers intensified during this era, setting the stage for future competitions in design, technology, and marketing. Their experiences during these challenging times contributed to shaping strategies that would influence the automotive industry for decades to come.

As we proceed to the next chapters, we will see how Ford and Chevrolet continued to evolve, how they influenced American culture, and how their rivalry pushed the boundaries of what automobiles could offer. The lessons from the 1920s and 1930s would inform their approaches in the years leading up to and following World War II, further solidifying their roles as icons of the automotive world.

Chapter 3: Post-War Prosperity and the Rise of the American Dream

The thunderous echoes of World War II had finally faded, leaving behind a world eager for peace, normalcy, and progress. In the United States, the end of the war marked the beginning of an era characterized by unprecedented economic growth and social change. Soldiers returned home to a nation poised for prosperity, and with them came a surge in consumer demand that would reshape industries across the board. At the forefront of this transformation stood the automotive giants, Ford and Chevrolet, ready to cater to the aspirations of a burgeoning middle class eager to embrace the American Dream.

The Post-War Economic Boom

The late 1940s and 1950s were a time of economic expansion in the United States. Government policies, including the G.I. Bill, provided veterans with opportunities for education, housing, and business ownership. This influx of educated and skilled workers fueled productivity and innovation. The construction of suburban homes skyrocketed, facilitated by affordable mortgages and the desire for a comfortable family life.

Consumer spending became the engine driving the economy. After years of wartime rationing and sacrifices, Americans were ready to indulge in goods and services that symbolized success and modernity. Automobiles, in particular, became emblematic of personal freedom, status, and the fulfillment of the American Dream.

Ford and Chevrolet: Meeting the Demand

Reconversion and Ramp-Up

Both Ford and Chevrolet faced the challenge of reconverting

their factories from wartime production back to civilian automobiles. During the war, automobile manufacturing had been repurposed to support the military effort, producing tanks, aircraft, and other equipment. The transition back required retooling factories, updating designs, and anticipating consumer preferences in a peacetime economy.

By 1946, both companies had resumed car production, initially offering models that were essentially pre-war designs due to time constraints. However, they quickly recognized the need for innovation to capture the attention of consumers hungry for new and exciting products.

Understanding the New Consumer

The post-war consumer was different from previous generations. The middle class was expanding, and with higher disposable incomes, people sought products that offered not just utility but also style, comfort, and modern conveniences. Automobiles became a reflection of personal identity and success.

Both Ford and Chevrolet invested heavily in market research to understand these evolving consumer desires. They aimed to produce vehicles that embodied the optimism and forward-looking spirit of the era.

Advancements in Car Design

The Influence of Aerodynamics and Jet Age Aesthetics

The 1950s saw the infusion of aerospace design elements into automobiles. The fascination with rockets and jets, spurred by advancements in aviation and the advent of the Space Race, influenced car styling. Tailfins, chrome accents, and streamlined bodies became popular, symbolizing speed and modernity.

Chevrolet embraced this trend with models like the 1957 Chevrolet Bel Air, which featured bold tailfins, a wide grille, and lavish use of chrome. The Bel Air became an icon of the era, representing the pinnacle of style and luxury for the average consumer.

Ford responded with designs like the 1955 Ford Fairlane, which showcased a sleek profile and innovative features. The Fairlane's Crown Victoria variant introduced the distinctive "basket handle" roof styling, setting it apart in a crowded market.

Color and Customization

Automakers began offering a wider palette of colors and customization options. Two-tone paint schemes, pastel colors, and personalized interiors allowed consumers to express individuality. This emphasis on aesthetics played a significant role in marketing and appealing to the desires of a style-conscious public.

Interior Comfort and Convenience

Advancements in interior design focused on comfort and convenience. Features such as adjustable seats, improved heating and ventilation systems, and enhanced sound insulation made driving more pleasurable.

Ford introduced the "Lifeguard" safety package in 1956, which included a padded dashboard, deep-dish steering wheel, and optional seat belts. This initiative highlighted a growing awareness of safety alongside comfort.

Chevrolet offered innovations like the "Powerglide" automatic transmission, power steering, and power brakes, making driving easier and more accessible to a broader audience.

Technological Innovations

Engine Performance and Efficiency

The post-war era witnessed significant advancements in engine technology. Consumers desired vehicles that not only looked good but also delivered impressive performance.

Ford developed the "Y-block" V8 engine, debuting in 1954, which offered improved horsepower and torque. This engine powered models like the Ford Thunderbird, introduced in 1955 as a "personal luxury car" combining performance with upscale features.

Chevrolet countered with the legendary "Small Block" V8 engine in 1955. Engineered by Ed Cole, this engine became renowned for its compact size, efficiency, and potential for performance enhancements. It powered vehicles like the Chevrolet Corvette, America's first mass-produced sports car introduced in 1953.

Automatic Transmissions and Power Accessories

Automatic transmissions became increasingly popular, reducing the complexity of driving and appealing to the growing number of female drivers and urban commuters.

Chevrolet's "Powerglide" and Ford's "Ford-O-Matic" transmissions offered smooth shifting and ease of use. Power accessories, including windows, seats, and convertible tops, added to the luxury and convenience of the vehicles.

Safety Features

While style and performance were paramount, safety began to emerge as a concern. Both companies experimented with safety features, though consumer interest was initially lukewarm.

Ford's "Lifeguard" safety campaign aimed to educate consumers about the importance of safety features. Although the campaign did not immediately translate into higher sales, it set the groundwork for future safety innovations.

Marketing to the Middle Class

The Power of Advertising

Advertising became a crucial tool in reaching consumers. Print ads, radio spots, and the burgeoning medium of television allowed automakers to showcase their latest models to a wide audience.

Chevrolet's advertising often emphasized the emotional appeal of car ownership—freedom, adventure, and family togetherness. Slogans like "See the USA in your Chevrolet" tapped into the nation's love for road trips and exploration.

Ford utilized celebrity endorsements and sponsored popular television programs like "The Ford Show," starring Tennessee Ernie Ford. These strategies increased brand visibility and associated Ford with quality entertainment.

Financing and Accessibility

To accommodate the financial needs of the middle class, both companies offered financing options that made car ownership more attainable. Low down payments and extended payment plans enabled consumers to purchase new cars without significant upfront costs.

Dealership networks expanded, providing accessible locations for sales, service, and community engagement. Showrooms became destinations where families could experience the latest models firsthand.

The Suburban Lifestyle and the Car Culture

Cars as Essential to Suburban Living

The post-war housing boom led to the development of suburbs, where owning a car became a necessity rather than a luxury. Commuting to urban centers for work, accessing shopping centers, and transporting children required reliable personal transportation.

Ford and Chevrolet recognized this shift and tailored their offerings to meet the practical needs of suburban families. Station wagons gained popularity, providing ample space for passengers and cargo.

Models like the Ford Country Squire and Chevrolet Nomad combined utility with style, featuring woodgrain paneling and upscale interiors. These vehicles became symbols of the ideal American family life.

The Rise of Car Culture

Automobiles began to permeate various aspects of American culture. Drive-in theaters, diners, and car shows became popular social venues. Teenagers embraced cars as a means of independence and social status.

Hot rodding and customizing cars emerged as subcultures. Enthusiasts modified vehicles for performance and aesthetics, often choosing models from Ford and Chevrolet due to their availability and the adaptability of their designs.

Music and movies celebrated car culture, with songs like "Hot Rod Lincoln" and films like "Rebel Without a Cause" highlighting the automobile's role in youth identity and rebellion.

Competition Intensifies

The Battle of the Brands

The rivalry between Ford and Chevrolet intensified during this period, each striving to outdo the other in design, technology, and market share.

In 1949, Ford introduced its first all-new post-war car, featuring modern slab-sided styling and integrated fenders. The "Shoebox Ford," as it was nicknamed, was a departure from pre-war designs and was well-received.

Chevrolet responded with the 1950 "Styleline" and "Fleetline" models, offering updated styling and improved features. The competition pushed both companies to continually innovate and refine their offerings.

Pricing Strategies

Competitive pricing remained essential. Both companies sought to provide value without sacrificing quality. They introduced models at various price points, ensuring that there was an option for consumers across different income levels.

Special promotions, trade-in allowances, and dealer incentives were employed to attract buyers. This fierce competition benefited consumers, who enjoyed a wider selection of affordable and feature-rich vehicles.

Impact on the Automotive Industry

Setting New Standards

The advancements made by Ford and Chevrolet during this era set new standards for the automotive industry. The emphasis on design, comfort, and technology became expectations for consumers.

Other automakers were compelled to follow suit or risk losing market share. The innovations introduced during this time laid the groundwork for future developments in safety, performance, and convenience.

Global Influence

The success of Ford and Chevrolet in the United States had global implications. Both companies expanded their international operations, exporting vehicles and establishing manufacturing facilities abroad.

Their designs and technologies influenced automotive trends worldwide. The concept of the American car became synonymous with style, power, and modernity on the global stage.

Challenges and Adjustments

Labor Relations and Production

The rapid expansion required managing labor relations carefully. Both companies faced strikes and demands for better wages and working conditions.

In 1945, Ford recognized the United Auto Workers (UAW) union, leading to negotiations that improved labor relations. Chevrolet, as part of GM, also worked with the UAW to address workers' concerns.

Automation began to enter the manufacturing process, increasing efficiency but also raising concerns about job security. Balancing technological advancements with workforce stability became an ongoing challenge.

Balancing Tradition and Innovation

While innovation was essential, both companies had to balance it with brand identity and consumer expectations. Radical changes risked alienating loyal customers, while stagnation could render models obsolete.

Market research and consumer feedback became integral to product development. Focus groups, surveys, and dealer input helped guide decisions, ensuring that new models aligned with market desires.

Embracing the Future

Introduction of New Models

The late 1950s saw the introduction of iconic models that would define the brands for years to come.

Ford launched the Edsel in 1958, aiming to fill a perceived gap between its Ford and Mercury lines. However, the Edsel became a case study in market miscalculation, failing to resonate with consumers due to its styling, pricing, and timing amid a recession.

Chevrolet, meanwhile, introduced the Impala in 1958, which became a best-seller due to its attractive design, performance options, and value. The Impala's success reinforced Chevrolet's understanding of consumer preferences.

Technological Milestones

Both companies continued to push technological boundaries. Fuel injection systems, improved suspension designs, and enhanced braking systems were introduced to improve performance and safety.

Concept cars and prototypes showcased visions of the future, including features like turbine engines, advanced aerodynamics, and novel materials. While not all concepts reached production, they demonstrated a commitment to innovation.

Cultural Impact and Legacy

Symbolism of the American Dream

The automobiles produced by Ford and Chevrolet during this era became symbols of the American Dream. They represented economic success, personal freedom, and the optimism of a nation confident in its future.

Ownership of a new car was a milestone for many families, marking progress and achievement. The vehicles became integral to family life, facilitating vacations, daily commutes, and social activities.

Influence on Design and Lifestyle

The styling cues and features introduced influenced other industries, including architecture, fashion, and product design. The emphasis on sleek lines, futuristic elements, and bold colors permeated the aesthetics of the time.

Automobiles shaped lifestyles, enabling the growth of suburbs, altering shopping habits with the rise of shopping centers, and changing leisure activities. The mobility they provided reshaped the social and physical landscape of the nation.

Conclusion: A Golden Era of Innovation

The post-war period was a golden era for Ford and Chevrolet, marked by rapid growth, fierce competition, and significant advancements in automotive technology and design. Both companies successfully tapped into the aspirations of the middle class, offering products that embodied the spirit of the times.

Their innovations not only satisfied consumer demands but also pushed the entire industry forward. The emphasis on style, comfort, and performance set new benchmarks and redefined what consumers expected from their vehicles.

The rivalry between Ford and Chevrolet during this era was a catalyst for progress, each pushing the other to achieve greater heights. The vehicles they produced became enduring icons, celebrated in popular culture and cherished by enthusiasts.

As the nation looked ahead to the 1960s, new challenges and opportunities awaited. The seeds planted during this period would give rise to the next phase of the rivalry, including the emergence of the muscle car era and the legendary battles between models like the Ford Mustang and Chevrolet Camaro.

The post-war prosperity and the rise of the American Dream had set the stage for continued innovation and competition, ensuring that the open road remained a place of excitement, freedom, and endless possibilities.

Chapter 4: Enter the Muscle Cars

The 1960s in America were a time of profound social change, marked by the rise of a youthful counterculture, economic prosperity, and an insatiable appetite for innovation. It was an era when the nation's roads became a stage for a new kind of performance—the muscle car. This movement didn't just introduce a new class of automobiles; it intensified the longstanding rivalry between Ford and Chevrolet, propelling it into high gear. The advent of muscle cars and the subsequent launch of the Ford Mustang and Chevrolet Camaro would forever alter the automotive landscape and embed these vehicles into the fabric of American culture.

The Birth of the Muscle Car Phenomenon

Cultural and Economic Backdrop

Post-World War II prosperity had given rise to a generation with disposable income and a desire for excitement. The Baby Boomers were coming of age, and their tastes differed markedly from those of their parents. They craved speed, style, and individuality. The car was more than transportation; it was a statement of identity.

At the same time, advancements in automotive technology made it feasible to produce powerful engines at a reasonable cost. The combination of a youthful market eager for performance and the ability to deliver it set the stage for the muscle car revolution.

Defining the Muscle Car

A muscle car is traditionally defined as a mid-sized car equipped with a large, powerful V8 engine, designed for high-performance driving. These cars were affordable and accessible, allowing ordinary consumers to experience the thrill of speed and power previously reserved for more expensive sports cars.

While various manufacturers contributed to the muscle car era, the rivalry between Ford and Chevrolet would bring the competition to its zenith, each pushing the other to create more powerful and desirable machines.

Early Performance Models Leading Up to the Mustang and Camaro

Ford's Foray into Performance

Before the Mustang's debut, Ford had already dipped its toes into high-performance waters. Models like the Ford Galaxie and the Fairlane were available with potent engines.

- **Ford Galaxie 500 XL**: Introduced in the early 1960s, the Galaxie 500 XL was a full-sized car that could be equipped with engines like the 390 cubic inch (6.4L) V8 and later the 427 cubic inch (7.0L) V8. These cars were successful in NASCAR racing, bolstering Ford's performance image.

- **Ford Fairlane Thunderbolt**: In 1964, Ford produced the Fairlane Thunderbolt, a limited-production drag racing car. It featured the powerful 427 V8 engine and was stripped down for maximum performance. The Thunderbolt dominated drag strips but was not intended for mass-market sales.

Chevrolet's Performance Offerings

Chevrolet was also active in the performance arena with models that appealed to enthusiasts seeking power and speed.

- **Chevrolet Impala SS**: The Super Sport (SS) package became available for the Impala in 1961. With options for high-performance engines like the 409 cubic inch (6.7L) V8, the Impala SS offered both style and muscle. The Beach Boys immortalized the "409" in their hit song, embedding it in popular culture.

- **Chevrolet Chevelle Malibu SS**: Introduced in 1964, the Chevelle was Chevrolet's entry into the mid-sized market. The Malibu SS trim offered sporty styling and performance options, laying the groundwork for more powerful iterations.

The Pontiac GTO Influence

While not a Chevrolet model, the Pontiac GTO, introduced in 1964 by GM's Pontiac division, played a crucial role in igniting the muscle car craze. The GTO packaged a large 389 cubic inch (6.4L) V8 engine into the mid-sized Tempest body. Its success demonstrated the market potential for such cars and influenced both Ford and Chevrolet to develop their own muscle machines.

The Launch of the Ford Mustang

A New Breed: The Pony Car

In April 1964, at the New York World's Fair, Ford unveiled the Mustang. This car created an entirely new segment—the "pony car." The term referred to affordable, compact, highly styled cars with a sporty or performance-oriented image.

- **Design and Concept**: The Mustang was based on the compact Ford Falcon platform but featured a long hood, short deck, and seating for four. Its design was both sporty and practical, appealing to a wide audience.

- **Customization and Affordability**: Starting at a base price of $2,368, the Mustang was affordable for young buyers. It offered a plethora of options, allowing customers to personalize their cars with choices in engines, colors, interiors, and accessories.

Market Impact

The Mustang was an instant success, far exceeding Ford's sales projections. Over 22,000 orders were placed on the first day. By the end of the first year, Ford had sold more than 400,000 Mustangs.

The car's popularity was fueled by its appearance in movies and television, such as the James Bond film "Goldfinger," and endorsements by celebrities. The Mustang became a symbol of youthful freedom and performance.

Performance Versions

While the base Mustang was equipped with modest engines, Ford soon introduced performance variants:

- **Mustang GT**: Offered upgraded suspension, disc brakes, and more powerful engines, including the 289 cubic inch (4.7L) V8.

- **Shelby GT350**: Developed in partnership with racing driver Carroll Shelby, the GT350 was a high-performance version designed for racing homologation. It featured a tuned 289 V8 engine producing 306 horsepower.

Chevrolet Responds with the Camaro

The Need for Competition

The Mustang's success caught Chevrolet off guard. Recognizing the threat, Chevrolet set out to develop a competitor that could challenge the Mustang's dominance in the pony car market.

Development of the Camaro

- **Project Designation**: Internally known as "Panther" during development, the car was officially named "Camaro" upon release. Chevrolet explained that Camaro was a French word meaning "comrade" or "friend," though the term was actually coined to fit Chevrolet's naming convention.

- **Launch and Features**: The Camaro debuted in September 1966 as a 1967 model. It shared the new F-body platform with the Pontiac Firebird. The Camaro featured a long hood, short rear deck, and was available in coupe and convertible body styles.

Customization and Performance

Like the Mustang, the Camaro offered extensive customization:

- **Engine Options**: Ranged from a 230 cubic inch (3.8L) inline-six to powerful V8s like the 396 cubic inch (6.5L) engine.

- **Trim Levels**: Included the base model, RS (Rally Sport) with aesthetic upgrades, SS (Super Sport) with performance enhancements, and the Z/28, which was designed for Trans-Am racing homologation.

Market Reception

The Camaro was well-received, quickly becoming a serious contender to the Mustang. Chevrolet marketed the Camaro aggressively, emphasizing its performance capabilities and stylish design.

Intensifying the Rivalry

The Pony Car Wars

The introduction of the Camaro escalated the competition between Ford and Chevrolet, igniting what became known as the "Pony Car Wars." Both companies were locked in a battle to outdo each other in terms of performance, style, and innovation.

- **Advertising Battles**: Advertisements often took subtle jabs at the competition. Ford and Chevrolet highlighted their own strengths while pointing out the perceived weaknesses of the other.

- **Performance Showdowns**: Magazine comparisons, drag races, and motorsport events became arenas for proving superiority. Enthusiasts eagerly followed the latest developments and performance figures.

Expanding Model Lineups

Both companies expanded their offerings to cater to various segments of the market:

- **Ford's Expansions**:
    - **Boss 302 and Boss 429 Mustangs**: High-performance models designed to compete in racing and showcase Ford's engineering prowess.
    - **Mach 1**: Introduced in 1969, offered a blend of performance and style.

- **Chevrolet's Expansions**:
    - **Camaro Z/28**: Equipped with a high-revving

302 cubic inch (4.9L) V8, it was successful in Trans-Am racing.

- **Yenko Camaro**: A special dealer-modified Camaro equipped with a 427 cubic inch (7.0L) engine, offering extreme performance.

Impact on Motorsports

The rivalry extended to the racetrack, where success translated into sales:

- **Trans-Am Racing**: Both the Mustang and Camaro competed fiercely in the Trans-American Sedan Championship. Victories in this series boosted the performance image of the brands.

- **Drag Racing**: Factory-supported drag cars and privateer efforts showcased the raw power of these muscle cars. The NHRA (National Hot Rod Association) events were popular venues.

Cultural Influence

The muscle car rivalry permeated American culture:

- **Music and Media**: Songs like "Mustang Sally" by Wilson Pickett and "Camaro" by Kings of Leon highlighted the cars' iconic status.

- **Movies and Television**: Mustangs and Camaros appeared in films and shows, further embedding them in popular culture. Steve McQueen's 1968 film "Bullitt" featured a famous chase scene with a Mustang GT.

Technological Advancements and Innovations

Engine and Performance Technologies

Advancements in engine design and performance technologies were at the forefront:

- **Ford**:
    - **Cleveland and Windsor V8s**: New engine designs that offered improved performance.
    - **Suspension Enhancements**: Better handling through improved suspension geometries.
- **Chevrolet**:
    - **Big Block Engines**: The 396 and later 427 V8 engines provided substantial horsepower.
    - **Disc Brakes and Handling Packages**: Improved stopping power and handling to match the increased speed.

Safety and Emissions

As the decade progressed, concerns about safety and emissions began to influence design:

- **Federal Regulations**: The introduction of safety standards and emission controls required manufacturers to adapt. Both companies worked to comply without compromising performance.

Challenges and the Evolution of the Muscle Car

Insurance and Fuel Costs

The high-performance muscle cars began to face challenges:

- **Insurance Rates**: Insurance companies raised premiums on muscle cars due to increased accident rates among young drivers.

- **Fuel Consumption**: The powerful engines consumed significant amounts of fuel, which became a concern as fuel prices rose.

Emissions Regulations

The implementation of stricter emission standards in the early 1970s forced manufacturers to detune engines, leading to a decline in horsepower ratings.

- **Compression Ratios Lowered**: To meet regulations, compression ratios were reduced, impacting performance.

- **Shift Towards Efficiency**: Manufacturers began to focus on fuel efficiency and cleaner emissions.

Legacy of the Muscle Car Era

An Enduring Impact

The muscle car era left an indelible mark on the automotive world:

- **Cultural Iconography**: The Mustang and Camaro became symbols of American automotive culture.

- **Collector's Items**: Classic models from the 1960s are

highly sought after by collectors and enthusiasts.

- **Revival and Continuation**: Both Ford and Chevrolet have continued to produce modern versions of the Mustang and Camaro, paying homage to their heritage while incorporating contemporary technology.

Influence on Future Designs

The intense competition pushed both companies to innovate continually:

- **Performance Expectations**: Consumers came to expect high performance from American cars, influencing future designs.

- **Design Language**: The aesthetic principles established during this era have influenced automotive design for decades.

Conclusion: The Muscle Car Era Intensifies the Rivalry

The introduction of muscle cars in the 1960s significantly intensified the rivalry between Ford and Chevrolet. The Mustang and Camaro became the flagships of this competition, each embodying their manufacturer's strengths and philosophies.

This period was marked by rapid innovation, bold design choices, and a focus on delivering unparalleled performance to the average consumer. The rivalry fueled advancements that benefited car enthusiasts and solidified both brands' positions in automotive history.

As the 1960s drew to a close, the muscle car era faced new challenges, but the legacy of the Mustang and Camaro would endure. The next chapters will explore how this rivalry continued to evolve, impacting motorsports, facing regulatory changes, and adapting to shifting consumer preferences, all while maintaining the spirit of competition that began with the pioneers of the open road.

Chapter 5: The Ford Mustang—A New Breed of Pony Car

In the spring of 1964, a revolution rolled onto the stage of the New York World's Fair, cloaked in sleek lines and powered by the ambitions of a generation. The Ford Mustang wasn't just another car—it was a bold statement, a reflection of youthful exuberance, and the embodiment of the American spirit during a time of significant social change. This chapter delves into the conception, development, and immediate impact of the Mustang, exploring how it became an instant classic and a cultural icon that would forever alter the automotive landscape.

The Genesis of an Idea

Market Shifts and Consumer Desires

The early 1960s were marked by a shifting demographic landscape in the United States. The Baby Boom generation was reaching driving age, and there was a burgeoning market of young consumers with disposable income. These potential buyers sought vehicles that reflected their aspirations—cars that were affordable yet stylish, practical yet exciting.

Ford Motor Company recognized this untapped market segment. Lee Iacocca, then Vice President and General Manager of the Ford Division, was a key proponent of developing a new kind of car. He envisioned a vehicle that would appeal to young buyers and compete with sporty European imports that were gaining popularity.

Project T-5: From Concept to Creation

The project that would become the Mustang was internally known as T-5. The development team faced the challenge of creating a car that was both cost-effective and compelling. To achieve affordability, they decided to base the new car on the existing platform of the Ford Falcon, a compact and economical model.

Designer Gale Halderman and his team were tasked with crafting a vehicle that exuded sportiness and appeal. The result was a car with a long hood, short rear deck, and an aggressive stance. The Mustang's styling drew inspiration from European sports cars while maintaining a distinctly American identity.

Key design features included:

- **Iconic Front End**: The galloping horse emblem and the distinctive grille set the Mustang apart.

- **Side Scoops and Sculpted Lines**: These elements added to the car's dynamic appearance.

- **Customizable Options**: The Mustang was designed to be highly customizable, offering a range of engines, colors, and accessories.

The Launch: A Phenomenon Unveiled

Debut at the New York World's Fair

On April 17, 1964, the Mustang made its public debut at the New York World's Fair. Simultaneously, Ford launched an unprecedented advertising blitz, including commercials broadcast on all three major television networks—ABC, CBS, and NBC—at the same time.

The response was immediate and overwhelming. Showrooms were flooded with eager buyers. Some dealerships reported selling out of their initial stock within hours. Ford received more than 22,000 orders on the first day alone.

Specifications and Pricing

The Mustang's base model came with a 170 cubic inch (2.8L) inline-six engine, priced at an accessible $2,368. However, Ford offered a range of options that allowed buyers to tailor the car to their preferences:

- **Engine Choices**: Options included the 260 cubic inch (4.3L) V8 and the more powerful 289 cubic inch (4.7L) V8.

- **Body Styles**: Initially available as a hardtop and convertible, with the fastback joining later in 1965.

- **Transmission Options**: Offered both manual and automatic transmissions.

- **Customization**: Over 70 factory options and numerous dealer-installed accessories.

This level of customization was revolutionary, enabling buyers to create anything from an economical runabout to a high-performance sports car.

Marketing Mastery

Ford's marketing strategy was integral to the Mustang's success. The company positioned the car as a symbol of freedom and adventure. Advertisements featured youthful, energetic imagery, appealing to the aspirations of the target demographic.

Taglines like "The car to be designed by you" emphasized personalization. The Mustang was not just a car; it was an extension of the owner's personality.

Immediate Impact on the Automotive Market

Sales Success

The Mustang's sales exceeded all expectations. Ford had projected first-year sales of around 100,000 units. Instead, they sold over 400,000 Mustangs in the first year and surpassed one million units within two years.

This unprecedented success forced competitors to take notice. The Mustang had effectively created a new market segment—the pony car—and other manufacturers scrambled to develop their own offerings.

Industry Influence

The Mustang's impact on the automotive industry was profound:

- **Competitive Response**: General Motors accelerated the development of the Chevrolet Camaro, while other manufacturers introduced models like the Pontiac Firebird, Dodge Challenger, and Plymouth Barracuda.

- **Shift in Design Philosophy**: The success of the Mustang demonstrated the viability of combining sporty aesthetics with mass-market appeal.

- **Emphasis on Youth Market**: Automakers began to prioritize younger consumers in their product planning and marketing efforts.

Capturing the Spirit of the Times

Reflection of Cultural Shifts

The 1960s were a period of significant social and cultural change. The civil rights movement, the space race, and the rise of rock and roll music all contributed to an atmosphere of transformation. The Mustang embodied this spirit:

- **Symbol of Freedom**: The car's name and emblem—a wild, running horse—evoked images of unbridled freedom and the open road.

- **Youthful Energy**: The Mustang was marketed as the car for the young and the young at heart, aligning with the rise of youth culture.

- **Affordability and Accessibility**: By making the car attainable, Ford tapped into the aspirations of a generation seeking independence and self-expression.

Influence on Popular Culture

The Mustang quickly became ingrained in American popular culture:

- **Film and Television**: The Mustang made notable appearances in movies and TV shows, enhancing its iconic status. Notably, a 1968 Mustang GT 390 starred in the film "Bullitt," featuring Steve McQueen in one of cinema's most famous car chase scenes.

- **Music**: Songs like Wilson Pickett's "Mustang Sally" celebrated the car, further cementing its place in cultural consciousness.

- **Celebrity Ownership**: High-profile figures driving Mustangs added to the car's allure.

Evolution and Performance Enhancements

High-Performance Variants

Recognizing the appetite for performance, Ford introduced several high-performance versions of the Mustang:

- **GT Package**: Launched in 1965, the GT package included a more powerful engine, upgraded suspension, and disc brakes.

- **Shelby Mustang GT350**: In collaboration with Carroll Shelby, Ford produced the GT350, a race-ready Mustang with a tuned 289 V8 engine producing over 300 horsepower. The GT350 dominated SCCA (Sports Car Club of America) racing circuits.

- **Shelby GT500**: Introduced in 1967, featuring a 428 cubic inch (7.0L) V8 engine, the GT500 offered even more power and performance.

Design Refinements

The Mustang underwent several design updates in its early years:

- **1965**: Introduction of the fastback body style, adding a sleeker profile.

- **1967-1968**: The Mustang grew slightly in size to accommodate larger engines. Styling became more aggressive, with a wider grille and revised tail lights.

- **1969-1970**: Further styling changes included quad headlights and a more muscular appearance.

These evolutions kept the Mustang fresh and competitive in a rapidly changing market.

Challenges and Adaptations

Balancing Mass Appeal and Performance

One of the key challenges for Ford was maintaining the Mustang's broad appeal while catering to enthusiasts seeking higher performance:

- **Variety of Options**: Ford continued to offer a wide range of options to suit different buyers, from the economical six-cylinder models to the high-powered V8s.

- **Marketing Strategies**: Advertisements highlighted both the practicality and the excitement of owning a Mustang.

Competition Intensifies

As competitors introduced their own pony cars, Ford faced increased pressure:

- **Chevrolet Camaro**: Launched in 1966 as a direct competitor, the Camaro intensified the rivalry.

- **Plymouth Barracuda and Dodge Challenger**: Other manufacturers entered the fray, offering alternatives to the Mustang.

Ford responded by enhancing the Mustang's performance credentials and expanding its lineup to include special editions and limited-run models.

The Mustang's Enduring Legacy

Cultural Icon Status

The Mustang's impact transcended the automotive world:

- **Symbol of American Automotive Excellence**: The Mustang became synonymous with American cars, representing innovation and style.

- **Global Influence**: The car's popularity extended beyond the United States, influencing automotive design and culture worldwide.

- **Collector's Item**: Early Mustangs, particularly high-performance models like the Shelby GT350 and GT500, became highly sought after by collectors.

Continuation Through the Decades

The Mustang has remained in continuous production since its introduction, a testament to its enduring appeal:

- **Adaptation to Market Changes**: The Mustang has evolved to meet changing consumer preferences, incorporating new technologies and design trends.

- **Modern Performance**: Contemporary Mustangs offer advanced engineering, with models like the Shelby GT500 delivering over 700 horsepower.

- **Heritage and Nostalgia**: The Mustang continues to honor its roots, blending heritage styling cues with modern innovation.

Conclusion: An Instant Classic and Beyond

The Ford Mustang's development and launch in 1964 marked a pivotal moment in automotive history. By capturing the spirit of the times, Ford created a car that resonated deeply with a generation eager for change and excitement.

The Mustang's immediate success was the result of a confluence of factors:

- **Innovative Design**: Striking a balance between sportiness and practicality.

- **Marketing Savvy**: Understanding and appealing to the desires of young consumers.

- **Customization and Accessibility**: Offering a personalized experience at an affordable price.

The Mustang didn't just fill a market niche; it created one. Its influence prompted competitors to develop their own pony cars, fueling a new era of automotive competition.

Moreover, the Mustang became a cultural icon, symbolizing freedom, ambition, and the quintessential American love affair with the open road. Its legacy endures, not only in the continued production of new models but also in its lasting impact on popular culture and automotive design.

As we continue to explore the rivalry between Ford and Chevrolet, the Mustang stands as a shining example of how innovation, understanding of consumer desires, and strategic marketing can combine to create a product that defines an era. The Mustang's story is not just about a car; it's about a moment in time when ambition met opportunity, resulting in a legend that continues to gallop into the future.

Chapter 6: Chevrolet Strikes Back—The Birth of the Camaro

The mid-1960s were a time of rapid change and fierce competition in the American automotive industry. Ford had struck gold with the introduction of the Mustang in 1964, creating a new market segment known as the pony car. The Mustang's unprecedented success caught many by surprise, including Chevrolet, which had been a dominant force in the industry for decades. Recognizing the urgent need to respond, Chevrolet embarked on a mission to develop a car that would not only compete with the Mustang but also surpass it. This chapter delves into the birth of the Chevrolet Camaro in 1966, examining its design, performance, and the strategic moves Chevrolet made to reclaim its position in the market.

The Wake-Up Call: Mustang's Success

Ford's Market Disruption

The Mustang's launch was a game-changer. Ford had tapped into a desire among young consumers for a stylish, affordable, and sporty car. With over 400,000 units sold in its first year, the Mustang wasn't just a new model—it was a phenomenon. The success of the Mustang highlighted a gap in Chevrolet's lineup and posed a direct challenge to General Motors (GM), Chevrolet's parent company.

Chevrolet's Initial Hesitation

Initially, Chevrolet was skeptical about the Mustang's potential impact. Some executives believed the surge in Mustang sales was a temporary trend that would fade. However, as the Mustang continued to dominate, it became clear that Chevrolet needed to act swiftly. The company's reputation and market share were at stake, and the pressure was mounting to deliver a compelling response.

Project XP-836: Conceiving the Camaro

Development Begins

In late 1964, GM authorized Chevrolet to develop a new car under the codename "XP-836." The objective was clear: create a vehicle that could compete head-to-head with the Mustang. Chevrolet aimed to produce a car that offered similar affordability and style but with superior performance and features.

Design Philosophy

Chevrolet's design team, led by Bill Mitchell and stylist Irvin Rybicki, set out to craft a car that embodied a bold and aggressive aesthetic. They wanted the new model to have a distinct identity while incorporating Chevrolet's design language.

Key design goals included:

- **Sleek Profile**: A long hood and short deck to convey speed and agility.

- **Coke-Bottle Styling**: Curved body sides resembling the shape of a classic Coke bottle, adding dynamism to the car's appearance.

- **Versatility**: The ability to accommodate a variety of engines and options to appeal to a broad audience.

Engineering Challenges

Engineering the new car presented several challenges:

- **Platform Selection**: Chevrolet needed a suitable platform that could be developed quickly. They settled on a modified version of the GM "F-body" platform, which would also be used for the Pontiac Firebird.

- **Performance Focus**: The car had to offer robust performance to attract enthusiasts. This meant designing a chassis capable of handling powerful engines.

- **Cost Constraints**: To compete with the Mustang's pricing, Chevrolet had to keep production costs in check without sacrificing quality.

The Naming Mystery: From Panther to Camaro

What's in a Name?

Throughout its development, the car was referred to internally as the "Panther." As the launch date approached, Chevrolet sought a name that would resonate with the public and fit within their naming conventions, which often started with the letter "C" (e.g., Corvette, Corvair, Chevelle).

Choosing "Camaro"

After considering over 2,000 names, Chevrolet settled on "Camaro." The word had no specific meaning but was said to be derived from a French slang term for "friend" or "companion." This aligned with the idea of the car being a close companion to the driver.

When asked, Chevrolet's marketing team played into the mystery. They quipped that a Camaro was "a small, vicious animal that eats Mustangs," a playful jab at their competitor.

The Launch of the Camaro

Unveiling to the Public

On September 12, 1966, Chevrolet officially introduced the Camaro to the world. The launch was orchestrated with significant media coverage, including a live press conference broadcast simultaneously in 14 cities—a first in automotive history.

Model Variants

The 1967 Camaro was available in multiple configurations:

- **Body Styles**: Offered as a two-door hardtop coupe or convertible.

- **Trim Levels**:

    - **Base Model**: Equipped with a 230 cubic inch (3.8L) inline-six engine.

    - **RS (Rally Sport)**: Featured hideaway headlights, revised taillights, and additional trim.

    - **SS (Super Sport)**: Came with a 350 cubic inch (5.7L) V8 engine, unique hood with simulated air intakes, and performance suspension.

    - **Z/28**: A special performance package designed for racing homologation, equipped with a 302 cubic inch (4.9L) V8 engine.

Customization Options

Chevrolet embraced the strategy of offering extensive customization:

- **Engine Choices**: Ranged from economical six-cylinder engines to powerful V8s, including the 396 cubic inch (6.5L) big-block.

- **Transmission Options**: Available with three-speed or four-speed manual transmissions and two-speed or three-speed automatic transmissions.

- **Interior Packages**: Offered various seating materials, colors, and convenience features.

- **Performance Upgrades**: Options included power steering, power brakes, and heavy-duty suspensions.

This approach allowed buyers to tailor the Camaro to their preferences, whether they desired a daily driver or a track-ready machine.

Design and Performance Comparison with the Mustang

Exterior Styling

Camaro

- **Aggressive Stance**: The Camaro featured a low, wide stance with a sculpted body that exuded a sense of motion even when stationary.

- **Front Fascia**: The RS package's hideaway headlights gave the Camaro a sleek and modern look.

- **Coke-Bottle Shape**: The distinctive curves set it apart from the Mustang's more conservative lines.

Mustang

- **Classic Lines**: The Mustang maintained its iconic long hood and short deck but had evolved to be slightly larger by 1967.

- **Styling Updates**: Ford had made incremental changes, adding a more muscular appearance but retaining the original design cues.

Interior Comfort and Features

Camaro

- **Driver-Oriented Cockpit**: The dashboard was designed to be functional and accessible, with an emphasis on the driver's experience.

- **Comfort Options**: Offered amenities like air conditioning, upgraded sound systems, and custom upholstery.

Mustang

- **Refined Interior**: The Mustang's interior was known for its quality materials and attention to detail.

- **Convenience Features**: Similar to the Camaro, it offered various comfort options to enhance the driving experience.

Performance Capabilities

Engine Performance

- **Camaro Engines**:
    - **Base Engine**: The inline-six provided adequate power for everyday use.
    - **V8 Options**: The 350 V8 in the SS model produced 295 horsepower. The 396 big-block

offered up to 375 horsepower.

- ○ **Z/28 Engine**: The high-revving 302 V8 was designed for road racing, producing around 290 horsepower but capable of much more with modifications.

- **Mustang Engines**:

    - ○ **Base Engine**: Also started with an inline-six.

    - ○ **V8 Options**: The 289 V8 was available in various states of tune, with the high-performance version producing 271 horsepower.

    - ○ **Shelby Mustangs**: The GT350 and GT500 offered even higher performance but were limited in production and higher in price.

Handling and Suspension

- **Camaro**:

    - ○ **Front Suspension**: Featured independent front suspension with coil springs.

    - ○ **Rear Suspension**: Used a traditional leaf-spring setup.

    - ○ **Handling Packages**: The Z/28 included stiffer springs, heavy-duty shocks, and larger anti-roll bars, improving cornering abilities.

- **Mustang**:

    - ○ **Suspension Setup**: Similar to the Camaro, with independent front suspension and a live rear axle with leaf springs.

- **Handling Characteristics**: The Mustang was praised for its balance but was generally considered less agile than the Camaro in stock form.

Technological Innovations

- **Camaro**:

  - **Disc Brakes**: Optional front disc brakes improved stopping power.

  - **Positraction Differential**: Enhanced traction during acceleration.

- **Mustang**:

  - **Safety Features**: Ford emphasized safety with options like front disc brakes and seat belts.

  - **Convenience Technology**: Offered features like intermittent wipers and tilting steering wheels.

Chevrolet's Strategic Positioning

Marketing Efforts

Chevrolet's marketing campaigns were aggressive and targeted:

- **Taglines**: Slogans like "The Hugger" highlighted the Camaro's handling prowess.

- **Advertising**: Used print, television, and radio to promote the Camaro's performance and style.

- **Comparison with Mustang**: While avoiding direct mentions, advertisements subtly emphasized areas where the Camaro excelled over the Mustang.

Racing Involvement

Chevrolet understood the importance of motorsports in building a performance image:

- **Trans-Am Racing**: The Camaro Z/28 was developed specifically for Trans-American Sedan Championship racing. Success on the track translated to showroom credibility.

- **Drag Racing**: Chevrolet supported drag racers with performance parts and technical assistance, showcasing the Camaro's straight-line speed.

Impact on the Market

Sales Performance

The Camaro made a significant impact in its first year:

- **Sales Figures**: Sold over 220,000 units in 1967, a strong showing for a new model.

- **Market Share**: While the Mustang still led in sales, the Camaro had successfully captured a substantial portion of the pony car market.

Competitive Pressure

The introduction of the Camaro intensified the competition:

- **Ford's Response**: Ford made enhancements to the Mustang, including introducing the 390 cubic inch (6.4L) V8 to keep pace with the Camaro's performance offerings.

- **Market Dynamics**: Other manufacturers accelerated development of their own pony cars, leading to a crowded and highly competitive segment.

The Camaro's Cultural Significance

Embracing the Youth Market

The Camaro resonated with young buyers:

- **Image of Rebellion**: Positioned as a car for those who wanted to stand out and make a statement.
- **Customization Culture**: The extensive options allowed owners to personalize their Camaros, fostering a strong sense of ownership and identity.

Media and Pop Culture

The Camaro quickly became a fixture in popular media:

- **Film and Television**: Appeared in movies and TV shows, often associated with speed and excitement.
- **Music and Art**: Featured in album covers, songs, and artwork that celebrated the car's style and performance.

Chevrolet's Continuous Improvement

Refinements and Special Editions

Chevrolet continued to evolve the Camaro:

- **1968 Updates**: Minor styling changes, improved suspension components, and the introduction of the

SS396 with even more power.

- **COPO Camaros**: Through the Central Office Production Order (COPO) system, dealers like Yenko Chevrolet ordered Camaros with the 427 cubic inch (7.0L) V8, creating some of the most powerful Camaros of the era.

- **1969 Redesign**: A more aggressive front end, updated interiors, and the addition of the cowl induction hood improved performance and aesthetics.

Innovation in Engineering

- **Fuel Injection Experiments**: Chevrolet explored advanced technologies, although widespread adoption would come later.

- **Safety Features**: Implemented features like energy-absorbing steering columns and improved crash protection.

Comparative Analysis: Aiming to Outperform the Mustang

Performance Edge

The Camaro aimed to outperform the Mustang in several key areas:

- **Engine Power**: With larger engine options, the Camaro often had an advantage in horsepower and torque.

- **Handling Dynamics**: The Camaro's suspension tuning and chassis design provided better handling characteristics, especially in the performance-oriented Z/28 model.

- **Racing Success**: Victories in Trans-Am racing bolstered the Camaro's performance credentials.

Design and Appeal

- **Modern Styling**: The Camaro's fresh design appealed to buyers seeking something different from the established Mustang.

- **Customization**: Chevrolet's broad range of options allowed customers to create a car that matched or exceeded the Mustang in features and performance.

Value Proposition

- **Competitive Pricing**: The Camaro was priced to compete directly with the Mustang, offering similar or better features for the cost.

- **Dealer Support**: Chevrolet's extensive dealer network and willingness to accommodate special orders enhanced customer satisfaction.

Challenges and Adaptations

Regulatory Pressures

As the late 1960s approached, new safety and emission regulations began to impact car design:

- **Compliance Efforts**: Chevrolet worked to meet standards without compromising performance.

- **Impact on Performance**: Increasing regulations would eventually lead to reductions in horsepower and changes in engineering approaches.

Market Saturation

The growing number of competitors in the pony car segment intensified competition:

- **Differentiation**: Chevrolet had to continually innovate to keep the Camaro distinct and appealing.

- **Consumer Preferences**: Shifts in consumer tastes required adjustments in marketing and product offerings.

Conclusion: Chevrolet's Successful Counterattack

The introduction of the Camaro represented a significant strategic move by Chevrolet to counter the Mustang's dominance. By delivering a car that combined aggressive styling, robust performance, and extensive customization, Chevrolet effectively struck back and reignited the rivalry.

The Camaro's impact was multifaceted:

- **Reasserting Chevrolet's Market Position**: Demonstrated Chevrolet's ability to respond swiftly and effectively to market changes.

- **Elevating the Pony Car Segment**: Raised the stakes, pushing both Chevrolet and Ford to innovate and improve continually.

- **Cultural Integration**: Like the Mustang, the Camaro became an integral part of American automotive culture.

The Camaro didn't just compete with the Mustang—it challenged it on all fronts, from the showroom to the racetrack. This rivalry fueled advancements in design and engineering that benefited consumers and enriched the automotive landscape.

As the 1960s progressed, the battle between the Mustang and Camaro intensified, embodying the spirit of competition that had always driven Ford and Chevrolet. The next chapters will explore how this rivalry evolved, influencing not only the companies and their products but also the broader context of motorsports, culture, and industry trends.

Through the Camaro, Chevrolet not only struck back but also secured a lasting legacy. The story of its birth is a testament to innovation, determination, and the relentless pursuit of excellence that defines the enduring rivalry between these two automotive giants.

Chapter 7: The Pony Car Wars—Mustang vs. Camaro

The late 1960s and 1970s marked a golden era in American automotive history—a time when the roar of V8 engines echoed through the streets and the thrill of horsepower captivated the nation. At the heart of this excitement was a fierce rivalry between two iconic vehicles: the Ford Mustang and the Chevrolet Camaro. This chapter delves deep into the direct competition between these two titans, exploring their battles on the sales floor, in advertising campaigns, and in the hearts of the public. We'll also highlight the legacy models that emerged from this rivalry, such as the Mustang Shelby GT350 and the Camaro Z/28, which have become legends in their own right.

The Stage is Set: A Clash of Titans

The Birth of a Rivalry

By the mid-1960s, the Ford Mustang had already taken the automotive world by storm, creating the pony car segment and capturing the imagination of a generation. Chevrolet, recognizing the threat to its market share, responded with the introduction of the Camaro in 1966. This set the stage for an intense rivalry that would define the era.

Defining the Pony Car

Pony cars were characterized by their sporty styling, affordable pricing, and performance options. They were designed to appeal to young, performance-minded consumers who desired the thrill of a sports car without the high price tag. The Mustang and Camaro epitomized this segment, each offering a unique blend of style, power, and personalization.

The Late 1960s: The Battle Intensifies

Design Evolution and Performance Enhancements

Ford Mustang

- **1967-1968 Updates**: The Mustang underwent significant changes, growing slightly in size to accommodate larger engines. The styling became more aggressive, with a wider grille and updated taillights.

- **Performance Options**: Ford introduced the 390 cubic inch (6.4L) V8 engine, offering up to 320 horsepower. The GT package continued to be popular among enthusiasts.

Chevrolet Camaro

- **1967-1968 Updates**: The Camaro maintained its sleek design but introduced minor styling tweaks. The RS, SS, and Z/28 packages offered varying levels of performance and aesthetics.

- **Performance Options**: The SS models could be equipped with the 396 cubic inch (6.5L) V8, delivering up to 375 horsepower. The Z/28, designed for Trans-Am racing, featured a high-revving 302 cubic inch (4.9L) V8.

Marketing Wars: Capturing the Public's Imagination

Ford's Strategy

- **Emotional Appeal**: Ford's advertising focused on the Mustang's freedom and excitement. Slogans like "Mustang—Only Mustang Makes It Happen!" emphasized the car's ability to fulfill dreams.

- **Celebrity Endorsements**: The Mustang was featured in movies and TV shows, enhancing its cool factor. Steve McQueen's "Bullitt" and its iconic chase scene solidified the Mustang's image as a performance car.

Chevrolet's Counterattack

- **Performance Emphasis**: Chevrolet highlighted the Camaro's power and handling. Advertisements used taglines like "The Hugger" to showcase the car's road-gripping capabilities.

- **Competitive Edge**: Some ads subtly targeted the Mustang, positioning the Camaro as the superior choice for those seeking real performance.

Public Reception and Sales Figures

- **Sales Competition**: The Mustang continued to outsell the Camaro in the late 1960s, but the gap was closing. In 1967, Ford sold approximately 472,000 Mustangs, while Chevrolet sold about 221,000 Camaros.

- **Market Impact**: The Camaro's entrance invigorated the pony car segment, pushing both companies to innovate and improve.

The Icons Emerge: Legacy Models

Ford Mustang Shelby GT350 and GT500

Collaboration with Carroll Shelby

- **GT350**: Introduced in 1965, the Shelby GT350 was a high-performance variant of the Mustang, featuring a modified 289 cubic inch (4.7L) V8 producing 306 horsepower.

- **GT500**: Launched in 1967, the GT500 came equipped with a 428 cubic inch (7.0L) V8, delivering around 355 horsepower.

Racing Success

- **Track Dominance**: The Shelby Mustangs were successful in SCCA racing, enhancing the Mustang's performance reputation.

- **Cultural Impact**: The Shelby models became symbols of American muscle and are highly sought after by collectors today.

Chevrolet Camaro Z/28 and COPO Models

Z/28: Born for the Track

- **Purpose-Built**: The Z/28 was designed to compete in the Trans-American Sedan Championship (Trans-Am). It featured a 302 cubic inch (4.9L) V8, specially engineered to meet the 5.0L displacement limit for the series.

- **Performance Enhancements**: Included heavy-duty suspension, power front disc brakes, and a Muncie 4-speed manual transmission.

COPO Camaros

- **Central Office Production Order (COPO)**: Allowed dealers to order Camaros with special equipment not available through regular channels.

- **Notable Models**:
    - **COPO 9560 and 9561**: Equipped with the all-aluminum 427 cubic inch (7.0L) ZL1 engine or

the iron-block L72 427 V8, producing well over 400 horsepower.

- ○ **Dealer Specials**: Dealers like Yenko Chevrolet created their own high-performance Camaros, further boosting the car's reputation.

The Early 1970s: A Shift in the Battlefield

Design Overhauls

1970 Mustang

- **Restyling**: The Mustang underwent a significant redesign, becoming larger and heavier. The new look featured a more pronounced grille and a bulkier appearance.

- **Engine Options**: Introduced the 429 cubic inch (7.0L) Cobra Jet and Super Cobra Jet engines, pushing performance to new heights.

1970 Camaro

- **Second Generation**: The Camaro received a complete makeover, adopting a European-inspired design with a longer hood and a fastback roofline.

- **Performance Focus**: Continued to offer potent engines like the 350 cubic inch (5.7L) V8 and the 396 (now actually 402 cubic inch) V8.

Advertising in a Changing Landscape

Ford's Messaging

- **Performance and Luxury**: Advertisements began to highlight not just performance but also comfort and luxury features, aiming to broaden the Mustang's appeal.

- **Safety Features**: Emphasized the inclusion of new safety technologies in response to growing consumer awareness.

Chevrolet's Approach

- **European Flair**: Marketed the new Camaro's styling and handling as having a European influence, appealing to consumers interested in sophistication.

- **Performance Credibility**: Continued to stress the Camaro's racing heritage and on-road capabilities.

Public Reception and Market Dynamics

- **Changing Preferences**: Consumers began to show interest in smaller, more fuel-efficient cars due to rising fuel prices and environmental concerns.

- **Sales Trends**: Mustang sales started to decline, while Camaro maintained steadier numbers, though the overall pony car market was shrinking.

Challenges of the 1970s: Regulation and Crisis

Impact of Emissions and Safety Regulations

- **Government Standards**: New regulations required reductions in emissions and improvements in safety features, leading to decreased horsepower and heavier vehicles.

- **Performance Decline**: Both Mustang and Camaro saw reductions in engine outputs, impacting their performance image.

Oil Crisis of 1973

- **Fuel Shortages**: The Arab oil embargo led to fuel shortages and skyrocketing gas prices.

- **Shift in Consumer Demand**: Buyers increasingly favored fuel-efficient vehicles, putting additional pressure on muscle and pony cars.

Adaptations by Ford and Chevrolet

Ford Mustang II

- **Downsizing**: In 1974, Ford introduced the Mustang II, a smaller and more fuel-efficient model based on the Pinto platform.

- **Public Reaction**: While criticized by enthusiasts for its lack of performance, the Mustang II sold well due to its practicality.

Chevrolet Camaro Continuation

- **Steady Course**: Chevrolet opted not to downsize the Camaro immediately but made efforts to improve fuel efficiency.

- **Performance Packages**: Reduced performance options but maintained the Camaro's sporty image.

Advertising Battles: Navigating a New Era

Shifting Messages

- **Ford**:
    - **Economy and Practicality**: Advertisements for the Mustang II emphasized fuel efficiency

and modern styling.

- **Heritage References**: Attempted to maintain a connection to the Mustang's performance roots despite the changes.

- **Chevrolet**:
    - **Value Proposition**: Highlighted the Camaro's style and affordability.
    - **Performance Nostalgia**: Leveraged the Camaro's past performance successes in marketing materials.

Sales Figures and Market Reception

- **Mustang II Success**: Despite mixed reviews, the Mustang II's sales were strong in its initial years, reflecting consumer priorities.

- **Camaro's Position**: The Camaro's sales remained relatively stable, but it began to lose ground as the market shifted.

Late 1970s: Revival Attempts

Ford Mustang Cobra II and King Cobra

- **Performance Appearance**: Introduced models with aggressive styling cues like hood scoops and racing stripes to recapture the performance image.

- **Limited Power**: Due to regulations, these models offered modest performance improvements but mostly focused on aesthetics.

Chevrolet Camaro Z28

- **Return of the Z28**: Reintroduced in 1977 after a brief hiatus, the Z28 focused on handling and styling.

- **Sales Boost**: The Z28's return helped increase Camaro sales, appealing to buyers seeking performance-oriented cars.

Advertising and Public Response

- **Ford's Strategy**: Marketed the special edition Mustangs as a blend of style and sportiness.

- **Chevrolet's Approach**: Emphasized the Z28's performance heritage and improved handling.

Legacy and Lasting Impact

Cultural Significance

- **Enduring Icons**: Both the Mustang and Camaro became symbols of American automotive culture, representing freedom, performance, and individuality.

- **Influence on Media**: Featured prominently in films, music, and television, they left an indelible mark on popular culture.

Collector's Treasures

- **Desirable Models**: Special editions like the Mustang Shelby GT350/GT500 and Camaro Z/28 are highly sought after by collectors.

- **Restorations and Modifications**: Enthusiasts continue to restore and modify these classics, keeping the legacy alive.

Revival in Later Decades

- **Mustang's Return to Roots**: In the 1980s and beyond, Ford reintroduced performance-oriented Mustangs, rekindling the original spirit.

- **Camaro's Evolution**: Chevrolet continued to evolve the Camaro, balancing modern demands with heritage styling.

Conclusion: A Rivalry That Shaped an Era

The pony car wars between the Mustang and Camaro during the late 1960s and 1970s were more than a battle between two cars—they were a reflection of the changing tides in American society, economy, and culture. This rivalry pushed both Ford and Chevrolet to innovate, adapt, and strive for excellence.

Key Takeaways

- **Innovation Driven by Competition**: The intense rivalry led to the creation of some of the most iconic cars in history, each pushing the boundaries of design and performance.

- **Marketing Mastery**: Both companies demonstrated the power of effective marketing in shaping public perception and driving sales.

- **Adaptation to Change**: The challenges of the 1970s tested both manufacturers, highlighting the importance of flexibility and understanding consumer needs.

The Legacy Lives On

The Mustang and Camaro continue to captivate enthusiasts and new generations of drivers. Their stories are interwoven with the fabric of American automotive history, symbolizing the enduring appeal of the open road and the thrill of driving.

As we move forward in our exploration of the Ford versus Chevrolet rivalry, we will see how these two giants navigated the subsequent decades, faced new challenges, and continued to influence the industry and culture. The pony car wars were a defining chapter, but the story is far from over.

Chapter 8: Battlegrounds Beyond the Showroom—Motorsports

The rivalry between Ford and Chevrolet was not confined to the streets and showrooms; it extended to the high-octane world of motorsports. From the thunderous roar of engines on NASCAR speedways to the explosive acceleration on drag strips, the competition between these automotive giants fueled legendary races, technological innovations, and a deep-seated passion among fans and drivers alike. This chapter explores how Ford and Chevrolet clashed on the racetrack, the key races that defined their rivalry, and how these battles spurred advancements that filtered down to consumer vehicles.

The Need for Speed: Motorsports as a Battleground

Motorsports and Brand Identity

Motorsports have long been a proving ground for automotive technology and a platform for manufacturers to showcase their prowess. Success on the racetrack translates into brand prestige, influencing consumer perceptions and driving sales. For Ford and Chevrolet, dominance in racing was a strategic imperative.

- **Marketing Impact**: "Win on Sunday, sell on Monday" became a mantra, reflecting how victories boosted showroom traffic.

- **Technological Testing**: Racing environments pushed vehicles to their limits, revealing weaknesses and opportunities for innovation.

Diverse Arenas of Competition

The rivalry manifested across various forms of motorsports:

- **NASCAR (National Association for Stock Car Auto Racing)**: The premier stock car racing series in the United States.

- **Drag Racing**: Quarter-mile sprints testing acceleration and speed.

- **Trans-Am Series**: Road racing series for production-based cars.

- **Endurance Racing**: Long-distance events like the 24 Hours of Le Mans.

NASCAR: The High-Speed Showdown

Early NASCAR Involvement

Ford's Presence

- **1949 Ford Coupe**: Ford's entry into NASCAR began in earnest with the 1949 Coupe, which became a favorite among drivers for its durability.

- **Factory Support**: Ford provided limited backing to teams, recognizing the promotional value of racing.

Chevrolet's Entry

- **1955 Chevrolet Bel Air**: The introduction of the small-block V8 made Chevrolet competitive.

- **Privateer Efforts**: Early Chevrolet participation was driven by independent teams using off-the-shelf parts.

The 1960s: Intensifying the Rivalry

Ford's Total Performance Program

- **Investment in Racing**: In 1962, Ford launched the "Total Performance" program, significantly increasing its involvement in various motorsports.

- **Successes**: Achieved victories in NASCAR, rallying, and endurance racing.

Chevrolet's Response

- **Internal Challenges**: GM instituted a ban on factory-backed racing in 1963, limiting Chevrolet's official participation.

- **Workarounds**: Despite the ban, Chevrolet continued to support teams indirectly through parts and engineering assistance.

Key NASCAR Races and Rivalries

1963 Daytona 500

- **Ford Dominance**: Ford driver Tiny Lund won the race, showcasing the effectiveness of Ford's investment.

- **Technological Edge**: The Ford Galaxie featured aerodynamic enhancements and powerful engines.

1967 and 1968 Seasons

- **Chevrolet's Return**: With the ban lifted, Chevrolet returned with the Chevy Chevelle.

- **Rivalry Heats Up**: Drivers like Bobby Allison (Ford)

and David Pearson (Ford) battled against Chevrolet's up-and-coming stars.

1970s Showdowns

- **Petty vs. Yarborough**: Richard Petty (Plymouth, later Ford) and Cale Yarborough (Chevrolet) engaged in fierce competition.

- **Innovation Arms Race**: Aerodynamic cars like the Ford Torino Talladega and Chevrolet Monte Carlo emerged.

Technological Advancements from NASCAR

- **Aerodynamics**: Lessons learned about airflow and drag reduction influenced production car designs.

- **Engine Development**: High-performance engines developed for racing were adapted for consumer models.

- **Safety Innovations**: Advances in roll cages, seat belts, and fuel systems improved driver safety and were applied to road cars.

Drag Racing: Quarter-Mile Conquests

The Rise of Drag Racing

- **Cultural Phenomenon**: Post-war America saw a surge in drag racing's popularity, especially among youth.

- **Manufacturer Involvement**: Ford and Chevrolet recognized the marketing potential and began producing factory drag cars.

Ford's Thunderbolt and Cobra Jet

Ford Fairlane Thunderbolt (1964)

- **Purpose-Built Racer**: Equipped with a 427 cubic inch (7.0L) V8, lightweight components, and stripped interiors.

- **Dominance**: The Thunderbolt was a force in NHRA Super Stock competition.

Mustang Cobra Jet (1968)

- **Engine Innovation**: The 428 Cobra Jet engine delivered exceptional power.

- **Success on the Strip**: Achieved victories in NHRA events, enhancing the Mustang's performance image.

Chevrolet's COPO Camaros and Novas

Central Office Production Order (COPO)

- **Bypassing Limitations**: Chevrolet dealers used the COPO system to order cars with high-performance engines not available in standard models.

COPO Camaro (1969)

- **Powerhouse Engine**: Featured the 427 cubic inch (7.0L) L72 V8.

- **Drag Strip Success**: The COPO Camaros were competitive in Stock and Super Stock classes.

Yenko and Baldwin-Motion

- **Dealer-Tuned Specials**: Dealers like Don Yenko

modified Camaros and Novas for drag racing.

- **Cult Following**: These cars became legends due to their rarity and performance.

Legendary Drag Races and Drivers

Bill "Grumpy" Jenkins

- **Chevrolet Icon**: Known for his tuning prowess and driving skills.

- **Innovations**: Pioneered small-block engine development and chassis tuning.

"Dyno" Don Nicholson

- **Ford Champion**: Excelled in various classes, driving cars like the Ford Fairlane and Mustang.

- **Contributions**: Advanced the use of lightweight materials and fuel injection.

Technological Contributions from Drag Racing

- **Engine Tuning Techniques**: Improved carburetion, ignition, and exhaust systems found their way into consumer models.

- **Transmission and Drivetrain**: Strengthened components developed for racing enhanced durability in production cars.

- **Lightweight Materials**: Use of aluminum and fiberglass to reduce weight influenced sports car design.

Trans-Am Series: Road Racing Rivalry

The Trans-American Sedan Championship

- **Overview**: Established in 1966 for production-based cars, promoting competition among manufacturers.

- **Classes**: Divided into over 2.0L and under 2.0L engine categories.

Ford Mustang's Early Success

Shelby Mustang GT350

- **Race-Ready**: Modified for competition, the GT350 secured victories in the series' early years.

- **Driver Achievements**: Jerry Titus won the 1967 championship driving a Mustang.

Chevrolet Camaro's Challenge

Z/28 Development

- **Homologation Special**: Created to meet Trans-Am regulations, featuring a 302 cubic inch (4.9L) V8.

- **Engineering Excellence**: Combined power with handling improvements.

Championship Battles

- **1968 and 1969 Seasons**: Mark Donohue, driving for Roger Penske's team, led the Camaro to consecutive championships.

- **Intense Rivalries**: Close races between Mustang and Camaro teams captivated fans.

Impact on Production Cars

- **Handling Improvements**: Suspension and brake advancements improved vehicle dynamics in consumer models.

- **Engine Performance**: High-revving, small-displacement V8s offered a balance of power and efficiency.

- **Styling Influences**: Racing liveries and aerodynamic aids inspired visual elements on production cars.

Endurance Racing: Ford's Quest for Glory

Ford vs. Ferrari at Le Mans

- **Background**: After a failed attempt to purchase Ferrari, Henry Ford II directed efforts to defeat Ferrari on the track.

- **GT40 Program**: Developed the Ford GT40, a purpose-built race car.

Le Mans Victories

- **1966 24 Hours of Le Mans**: Ford achieved a historic 1-2-3 finish, ending Ferrari's dominance.

- **Subsequent Wins**: Ford continued to win Le Mans in 1967, 1968, and 1969.

Technological Triumphs

- **Advanced Engineering**: Innovations in aerodynamics, engine performance, and reliability.

- **Legacy**: The GT40's success elevated Ford's global

performance image.

Chevrolet's Involvement

- **Corvette Racing**: Participated in endurance racing, though factory support was limited due to GM's racing policies.

- **Privateer Efforts**: Independent teams campaigned Corvettes with varying degrees of success.

Other Motorsport Arenas

Rally Racing

- **Ford's Participation**: The Ford Escort became a rally icon in Europe, though less relevant to the U.S. market.

- **Chevrolet's Limited Role**: Minimal involvement in rallying, focusing efforts elsewhere.

Off-Road Racing

- **Ford Bronco**: Achieved success in events like the Baja 1000, enhancing Ford's rugged image.

- **Chevrolet Trucks**: Competed in off-road racing, promoting durability and capability.

Technological Advancements Driven by Racing

Engine Development

- **High-Performance Engines**: Technologies like overhead camshafts, fuel injection, and turbocharging were refined.

- **Durability Improvements**: Enhanced materials and

construction techniques increased engine longevity.

Aerodynamics and Design

- **Wind Tunnel Testing**: Improved understanding of aerodynamics led to sleeker, more efficient car designs.

- **Downforce and Stability**: Spoilers and air dams developed for racing found applications in sports cars.

Safety Innovations

- **Driver Protection**: Roll cages, harnesses, and impact-absorbing structures improved safety.

- **Consumer Benefits**: Safety features trickled down to production vehicles, enhancing occupant protection.

Suspension and Handling

- **Advanced Suspensions**: Independent rear suspensions and adjustable components improved handling.

- **Tire Technology**: Development of high-performance tires enhanced grip and responsiveness.

Notable Drivers and Personalities

Ford Legends

- **AJ Foyt**: Achieved success across multiple disciplines, including winning Le Mans with Ford.

- **Parnelli Jones**: Drove the Boss 302 Mustang in Trans-Am, known for his aggressive style.

Chevrolet Icons

- **Dale Earnhardt Sr.**: "The Intimidator," a NASCAR legend driving Chevrolets, with seven championships.

- **Mark Donohue**: Renowned for his skill and engineering acumen, leading Camaro to Trans-Am victories.

The Fan Factor: Building Brand Loyalty

Passionate Fan Bases

- **Brand Allegiance**: Fans often identified strongly with either Ford or Chevrolet, fueling the rivalry.

- **Merchandising**: Sales of branded apparel, models, and memorabilia reinforced connections.

Community and Culture

- **Car Clubs and Events**: Gatherings like car shows and racing events fostered community among enthusiasts.

- **Media Coverage**: Magazines, television, and later the internet amplified the rivalry's visibility.

Legacy of the Motorsport Rivalry

Influence on Vehicle Development

- **Performance Models**: Racing success justified the production of high-performance street cars.

- **Innovation Adoption**: Technologies proven on the track accelerated their integration into consumer vehicles.

Economic Impact

- **Sales Boost**: Racing victories correlated with increased sales, benefiting both companies financially.

- **Aftermarket Growth**: Demand for performance parts and accessories expanded the automotive aftermarket industry.

Enduring Rivalry

- **Continued Competition**: The rivalry persists, with both brands actively participating in modern motorsports.

- **Cultural Significance**: The racing battles have become part of automotive folklore, celebrated by generations.

Conclusion: Racing Fuels the Rivalry and Innovation

The motorsport battles between Ford and Chevrolet extended their rivalry beyond the showroom, igniting competition that spurred technological advancements and captivated audiences worldwide. Racing provided a dynamic arena where engineering prowess, driver skill, and strategic innovation converged.

Key Takeaways

- **Technological Advancements**: Innovations developed for racing improved performance, safety, and efficiency in consumer vehicles.

- **Marketing Power**: Success on the track enhanced brand image and influenced consumer preferences.

- **Cultural Impact**: Legendary races and rivalries enriched the automotive culture, fostering a deep connection between brands and enthusiasts.

The Road Ahead

The legacy of these motorsport battles continues to influence both companies:

- **Modern Racing Programs**: Ford and Chevrolet remain active in NASCAR, IMSA, and other series.

- **Performance Vehicles**: The spirit of competition drives the development of modern performance cars like the Ford GT and Chevrolet Corvette C8.

- **Innovation Focus**: Racing remains a catalyst for exploring new technologies, including electrification and advanced materials.

As we move forward in our exploration of the Ford versus Chevrolet story, the impact of their motorsport rivalry is a testament to how competition can drive progress, inspire passion, and leave an indelible mark on history. The racetrack was more than a battleground; it was a crucible where innovation was forged, legends were made, and the spirit of rivalry was both challenged and celebrated.

Chapter 9: Trucks and Workhorses—F-Series vs. Silverado

In the vast landscape of American automotive history, few rivalries are as enduring and fiercely contested as that between Ford and Chevrolet in the truck segment. While the pony car wars between the Mustang and Camaro captured the imagination of young enthusiasts, it was in the rugged and utilitarian world of trucks where Ford and Chevrolet solidified their dominance, catering to a diverse clientele that ranged from hardworking farmers and tradespeople to adventurous outdoor enthusiasts and suburban families. This chapter delves into the competition between the Ford F-Series and the Chevrolet Silverado, exploring how these workhorses became central to both brands' identities and revenue streams, ultimately shaping the American truck market.

The Trucking Backbone: Understanding the Market

Evolution of the American Truck

The American truck market has its roots in the early 20th century, evolving from simple utility vehicles to sophisticated machines capable of handling a wide array of tasks. By the mid-20th century, trucks had become indispensable to industries, construction, agriculture, and everyday life. The demand for reliable, powerful, and versatile trucks set the stage for fierce competition among manufacturers, with Ford and Chevrolet emerging as the primary contenders.

Diverse Customer Base

Trucks serve a multifaceted customer base:

- **Commercial Use**: Businesses rely on trucks for transportation, construction, and various operational needs.

- **Personal Use**: Many consumers use trucks for towing, hauling, and recreational activities.

- **Off-Road and Adventure**: Enthusiasts seek trucks capable of handling rough terrains and outdoor adventures.

Understanding these varied needs was crucial for Ford and Chevrolet as they developed their truck lines to cater to different segments of the market.

Ford F-Series: The Legacy of Strength and Reliability

Origins and Evolution

Introduced in 1948, the Ford F-Series was designed to replace the wartime Ford F-1 truck. The F-Series quickly became synonymous with durability and performance, evolving through numerous generations to incorporate advancements in technology, design, and capability.

Generational Developments

First Generation (1948–1952)

- **Introduction**: The inaugural F-1 laid the foundation with its robust build and versatility.

- **Design**: Utilitarian and functional, focusing on reliability and ease of maintenance.

Second Generation (1953–1956)

- **Modernization**: Introduced a new chassis and updated styling.

- **Performance**: Enhanced engine options improved towing and payload capacities.

Third Generation (1957–1960)

- **Styling Changes**: Sleeker lines and more aerodynamic shapes.
- **Technology Integration**: Introduction of features like power steering and automatic transmissions.

Subsequent Generations (1961–Present)

- **Innovation**: Continuous improvements in engine performance, fuel efficiency, safety features, and interior comfort.
- **Versatility**: Expansion into various trims and configurations to meet diverse needs.

Iconic Models and Special Editions

Ford F-100 to F-150

- **F-100**: Became a favorite among small business owners and farmers for its reliability.
- **F-150**: Introduced in 1975, it quickly became the best-selling truck in the United States, embodying a balance of power, comfort, and technology.

Raptor and Super Duty Lines

- **F-150 Raptor**: Launched in 2010, designed for high-performance off-roading with enhanced suspension, powerful engines, and aggressive styling.
- **Super Duty (F-250, F-350)**: Catered to heavy-duty

commercial use, offering superior towing and payload capacities.

Technological Advancements

- **Engine Innovations**: From the V8s of the early days to modern EcoBoost engines, Ford has consistently pushed the envelope in power and efficiency.

- **Safety Features**: Introduction of advanced safety technologies such as airbags, traction control, and lane-keeping assist.

- **Connectivity and Comfort**: Integration of modern infotainment systems, ergonomic interiors, and driver-assist features.

Chevrolet Silverado: The Pinnacle of American Trucking

Introduction and Growth

Launched in 1998 as the successor to the Chevrolet C/K series, the Silverado was Chevrolet's answer to the growing dominance of the Ford F-Series. Designed to offer a blend of performance, comfort, and reliability, the Silverado quickly established itself as a formidable competitor in the truck market.

Generational Developments

First Generation (1999–2006)

- **Foundation**: Built on the GMT800 platform, it emphasized improved ride quality and towing capabilities.

- **Design**: Robust and muscular, with a focus on functionality and durability.

Second Generation (2007–2014)

- **Redesign**: Shifted to the GMT900 platform, featuring more refined styling and enhanced interior amenities.

- **Performance Enhancements**: Introduction of more powerful engine options and improved suspension systems.

Third Generation (2015–Present)

- **Modernization**: Adoption of the GMT K2XX platform, incorporating advanced technology and eco-friendly features.

- **Versatility**: Expanded lineup with various trims and configurations to cater to a broader audience.

Iconic Models and Special Editions

Chevrolet Silverado 1500, 2500HD, and 3500HD

- **1500**: The most popular variant, balancing capability with comfort for personal and light commercial use.

- **2500HD and 3500HD**: Designed for heavy-duty applications, offering maximum towing and payload capacities.

Trail Boss and High Country

- **Trail Boss**: Targeted at off-road enthusiasts with enhanced suspension, skid plates, and all-terrain tires.

- **High Country**: The premium trim level, featuring luxurious interiors, advanced technology, and superior performance.

Technological Advancements

- **Engine Technology**: From the robust V8s to the introduction of the Duramax diesel engines, Chevrolet has prioritized power and efficiency.

- **Safety and Driver Assistance**: Incorporation of features like adaptive cruise control, blind-spot monitoring, and automatic emergency braking.

- **Infotainment and Connectivity**: Advanced infotainment systems with smartphone integration, navigation, and premium sound options.

Market Dominance: Trucks as Brand Pillars

Revenue and Sales Figures

Trucks have become a cornerstone of both Ford and Chevrolet's revenue streams. The Ford F-Series and Chevrolet Silverado consistently vie for the title of America's best-selling truck, each maintaining a loyal customer base and attracting new buyers through continuous innovation and marketing.

- **Ford F-Series**: Maintains a slight edge in annual sales, often securing the position of the best-selling vehicle in the United States.

- **Chevrolet Silverado**: A close contender, with strong sales figures and market presence, especially in the medium and heavy-duty segments.

Brand Identity and Loyalty

Trucks embody the rugged, dependable image that both Ford and Chevrolet strive to project. For many consumers, owning a Ford F-Series or Chevrolet Silverado is a statement of reliability, strength, and capability.

- **Ford**: The F-Series is often associated with traditional American toughness and work ethic.

- **Chevrolet**: The Silverado emphasizes versatility, modernity, and a balance between performance and comfort.

Advertising and Marketing Strategies

Both brands invest heavily in advertising to reinforce their truck lines' strengths and appeal to their target demographics.

Ford's Approach

- **Emotional Storytelling**: Campaigns often highlight real-life stories of truck owners, showcasing how the F-Series supports their daily lives and businesses.

- **Performance Focus**: Emphasizing towing capacity, engine power, and advanced features to appeal to performance-oriented buyers.

- **Innovative Features**: Marketing the latest technological advancements, such as Pro Power Onboard and FordPass Connect, to attract tech-savvy consumers.

Chevrolet's Strategy

- **Versatility and Capability**: Highlighting the Silverado's ability to handle a wide range of tasks, from heavy-duty hauling to comfortable daily driving.

- **Modern Design**: Showcasing the truck's sleek design and advanced interiors to attract consumers looking for both style and functionality.

- **Heritage and Innovation**: Balancing the brand's rich history with its commitment to innovation, promoting the Silverado as a truck that honors tradition while embracing the future.

Dealer Networks and Customer Service

A robust dealer network is essential for maintaining sales and customer loyalty. Both Ford and Chevrolet have extensive dealership systems that provide not just sales but also maintenance, repair, and customization services.

- **Ford Dealerships**: Emphasize personalized service, performance packages, and aftersales support to enhance customer satisfaction.

- **Chevrolet Dealerships**: Focus on providing a comprehensive ownership experience, including financing options, certified pre-owned programs, and community engagement.

Community Engagement and Brand Ambassadors

Engaging with communities and leveraging brand ambassadors play a significant role in maintaining a strong market presence.

- **Ford**: Partners with organizations like the Ford Motor Company Fund to support community initiatives, education, and disaster relief.

- **Chevrolet**: Involved in sponsorships of sports events, music festivals, and automotive shows, reinforcing its image as a versatile and community-oriented brand.

Legacy Models: Showcasing Excellence

Ford F-150 Raptor and Lightning

- **F-150 Raptor**: Introduced in 2017, the Raptor is a high-performance variant designed for off-road adventures, featuring specialized suspension, powerful engines, and aggressive styling.

- **F-150 Lightning**: Ford's venture into electric trucks, combining traditional F-150 capabilities with modern electric technology, reflecting the brand's commitment to innovation and sustainability.

Chevrolet Silverado Trail Boss and Silverado EV

- **Silverado Trail Boss**: Launched in 2021, the Trail Boss caters to off-road enthusiasts with enhanced suspension, skid plates, and all-terrain tires.

- **Silverado EV**: Chevrolet's entry into the electric truck market, offering a zero-emission alternative without compromising on power and capability.

Special Editions and Collaborations

- **Ford Limited Editions**: Collaborations with designers and performance experts to create unique, limited-run models that appeal to collectors and enthusiasts.

- **Chevrolet Custom Camaros**: Partnerships with aftermarket companies to offer customized Silverado models with enhanced performance and aesthetic features.

Technological Innovations Driven by the Truck Battle

Engine and Performance Enhancements

The competition between the F-Series and Silverado has driven both brands to develop engines that deliver exceptional power, efficiency, and reliability.

- **Ford EcoBoost Engines**: Turbocharged engines that provide high power output with improved fuel efficiency, catering to both performance and environmental concerns.

- **Chevrolet Duramax Diesel Engines**: Known for their torque and towing capabilities, the Duramax engines are a staple in the Silverado lineup, especially for heavy-duty applications.

Advanced Safety and Driver Assistance

Both Ford and Chevrolet have integrated cutting-edge safety features to enhance driver and passenger protection.

- **Ford Co-Pilot360**: A suite of driver-assist technologies including automatic emergency braking, blind-spot monitoring, and adaptive cruise control.

- **Chevrolet Safety Assist**: Features like forward collision alert, lane keep assist, and automatic braking systems to improve overall vehicle safety.

Connectivity and Infotainment

Modern trucks are as much about connectivity and comfort as they are about performance and utility.

- **Ford SYNC and FordPass**: Seamless integration of smartphones, voice-activated controls, and remote vehicle management through the FordPass app.

- **Chevrolet MyLink**: An intuitive infotainment system with touchscreen controls, smartphone integration, and advanced navigation features.

Towing and Payload Innovations

Enhancements in towing and payload capacities have been a focal point in the truck rivalry, ensuring that both the F-Series and Silverado meet the demanding needs of their users.

- **Ford Pro Trailer Backup Assist**: Simplifies the process of backing up with trailers, enhancing usability and safety.

- **Chevrolet Trailering System**: Comprehensive towing packages that include trailer sway control, integrated trailer brake controllers, and advanced hitch systems.

Market Dominance and Revenue Impact

Sales Performance and Market Share

The Ford F-Series and Chevrolet Silverado consistently vie for the top spot in the American truck market. Their sales figures not only reflect their popularity but also their ability to adapt to changing market dynamics and consumer preferences.

- **Ford F-Series**: Frequently secures the title of the best-selling vehicle in the United States, with annual sales often exceeding one million units.

- **Chevrolet Silverado**: A strong contender, regularly ranking second with robust sales numbers, particularly excelling in the medium and heavy-duty segments.

Revenue Contributions

Trucks contribute significantly to the overall revenue of both Ford and Chevrolet, often accounting for a substantial portion of their automotive sales.

- **Ford**: The F-Series is a major revenue driver, with profits stemming from both sales and aftersales services such as parts and maintenance.

- **Chevrolet**: The Silverado similarly plays a critical role in Chevrolet's financial success, supported by strong demand across various truck categories.

Global Market Influence

While the F-Series and Silverado are primarily dominant in the North American market, their influence extends globally, with both brands exporting trucks and establishing manufacturing facilities abroad to cater to international demand.

- **Ford**: Expands its truck lineup in markets like Australia with the Ford Ranger, leveraging the F-Series' reputation for strength and reliability.

- **Chevrolet**: Offers trucks like the Chevrolet Colorado and Silverado in various global markets, adapting to regional preferences and regulations.

Brand Identity and Consumer Loyalty

Ford's Brand Image

The Ford F-Series embodies the brand's commitment to strength, reliability, and innovation. It appeals to a wide range of consumers, from commercial operators to recreational drivers, by offering versatile and dependable vehicles.

- **Reliability**: Known for longevity and durability, the F-Series is trusted by businesses and individuals alike.

- **Innovation**: Continual advancements in technology and performance keep the F-Series relevant and competitive.

Chevrolet's Brand Image

The Chevrolet Silverado represents versatility, modernity, and robust performance. It attracts consumers looking for a blend of power, comfort, and advanced features in their trucks.

- **Versatility**: The Silverado caters to diverse needs, from heavy-duty hauling to comfortable daily driving.

- **Modernity**: Emphasis on contemporary design and cutting-edge technology appeals to tech-savvy and style-conscious buyers.

Customer Loyalty and Community

Both Ford and Chevrolet have cultivated strong communities around their truck lines, fostering brand loyalty through exceptional customer experiences, community engagement, and ownership perks.

- **Ford Owners**: Benefit from a vast network of dealers, extensive warranty options, and exclusive Ford events.

- **Chevrolet Owners**: Enjoy access to specialized services, loyalty rewards, and active participation in Chevrolet-sponsored events and gatherings.

Challenges and Adaptations in the Truck Market

Environmental Regulations and Sustainability

The truck market faces increasing pressure to reduce emissions and improve fuel efficiency, prompting both Ford and Chevrolet to innovate towards more sustainable solutions.

- **Electric Trucks**: Ford has introduced the F-150 Lightning, an all-electric variant of its flagship truck, while Chevrolet is developing the Silverado EV, signaling a shift towards electrification.

- **Hybrid Technologies**: Incorporating hybrid powertrains to balance performance with environmental responsibility.

Technological Disruptions

Advancements in autonomous driving, connectivity, and advanced manufacturing techniques are reshaping the truck landscape.

- **Autonomous Features**: Integration of self-driving technologies to enhance safety and convenience.

- **Smart Manufacturing**: Adoption of Industry 4.0 practices to streamline production and improve quality control.

Economic Fluctuations and Consumer Trends

Economic shifts, such as fluctuations in fuel prices and changing consumer preferences, impact truck sales and market strategies.

- **Fuel Efficiency Demands**: Rising fuel costs drive the

need for more efficient engines and lighter materials.

- **Shift to Electric**: Growing environmental consciousness leads consumers to consider electric and hybrid trucks as viable alternatives.

Global Competition

Emerging competitors from both traditional automakers and new entrants in the electric vehicle market challenge Ford and Chevrolet's dominance.

- **Tesla Cybertruck**: Introduces a futuristic design and electric powertrain, appealing to a new segment of environmentally conscious and tech-savvy consumers.

- **Rivian R1T and GMC Hummer EV**: Offer alternative electric truck options, intensifying competition in the rapidly evolving market.

Future Outlook: Sustaining Dominance in a Changing World

Innovation and Adaptation

To maintain their leadership in the truck segment, Ford and Chevrolet must continue to innovate and adapt to emerging trends and technologies.

- **Electrification**: Expanding electric truck offerings to meet growing demand for sustainable vehicles.

- **Advanced Technologies**: Integrating AI, connectivity, and autonomous features to enhance the driving experience and safety.

Expanding Market Reach

Both brands are exploring opportunities to expand their market presence globally, tailoring their truck offerings to meet regional demands and preferences.

- **Global Models**: Adapting the F-Series and Silverado for international markets with specific features and compliance with local regulations.

- **Strategic Partnerships**: Collaborating with technology firms and other automakers to accelerate innovation and market penetration.

Sustainability Initiatives

Commitment to sustainability is becoming increasingly important, with both Ford and Chevrolet investing in environmentally friendly practices and materials.

- **Green Manufacturing**: Implementing sustainable manufacturing processes to reduce carbon footprints.

- **Recyclable Materials**: Utilizing recyclable and eco-friendly materials in truck construction to promote environmental responsibility.

Customer-Centric Strategies

Focusing on enhancing customer experiences through superior service, customization options, and community engagement will be crucial for sustained success.

- **Personalization**: Offering more extensive customization options to cater to individual preferences and needs.

- **Enhanced Service**: Providing exceptional aftersales

support, maintenance services, and ownership benefits to build long-term loyalty.

Conclusion: The Enduring Truck Rivalry

The competition between the Ford F-Series and Chevrolet Silverado has been a defining element of the American truck market for decades. This rivalry has driven both manufacturers to excel in design, performance, technology, and customer service, ensuring that consumers have access to some of the best trucks available. Trucks have become more than just work vehicles; they are symbols of strength, reliability, and versatility, integral to both brands' identities and financial success.

As the automotive landscape continues to evolve with advancements in technology, shifting consumer preferences, and increasing environmental concerns, the F-Series and Silverado remain at the forefront of innovation. Their ability to adapt and lead in the face of change will determine their continued dominance and legacy in the truck segment.

The story of the Ford F-Series and Chevrolet Silverado is a testament to the power of competition in driving progress and excellence. As these trucks continue to evolve, they carry with them the rich history and enduring spirit of two of America's most iconic automotive brands, ensuring that the truck rivalry between Ford and Chevrolet remains as robust and influential as ever.

Chapter 10: Navigating the Oil Crisis and Environmental Concerns

The 1970s ushered in a period of profound transformation and uncertainty for the global economy, with the automotive industry standing at the forefront of these changes. The decade was marred by two significant oil crises—in 1973 and 1979—that not only reshaped energy policies but also fundamentally altered consumer behavior and automotive manufacturing practices. For industry titans Ford and Chevrolet, these crises posed formidable challenges but also presented opportunities to innovate and adapt. This chapter examines the impact of the oil crises on both companies, exploring their strategic shifts toward fuel efficiency, the introduction of compact cars, and the myriad challenges they faced in navigating this turbulent era.

The 1970s Oil Crises: A Catalyst for Change

The First Oil Crisis: 1973

In October 1973, the Organization of Arab Petroleum Exporting Countries (OAPEC) proclaimed an oil embargo targeting nations perceived as supporting Israel during the Yom Kippur War. The embargo led to skyrocketing oil prices and severe fuel shortages, creating economic turmoil in the United States and other Western countries. Gasoline prices quadrupled, and long lines at gas stations became a common sight.

The Second Oil Crisis: 1979

The 1979 oil crisis, triggered by the Iranian Revolution, compounded the economic difficulties of the decade. Political instability in Iran disrupted oil production, leading to another surge in oil prices and further exacerbating fuel scarcity. The cumulative effect of these crises plunged the global economy into stagflation—a troubling combination of stagnant economic growth and high inflation.

Immediate Impact on the Automotive Industry

The sudden spike in oil prices and the subsequent fuel shortages had a profound impact on the automotive industry. Consumers, now acutely aware of the cost of fuel, began to prioritize fuel efficiency over horsepower and performance. This shift forced automakers to rethink their product lines and manufacturing strategies to align with the new market realities.

Ford and Chevrolet: Confronting the Crisis

Ford Motor Company: Adapting to a New Reality

Impact on Ford

Ford, traditionally known for its robust, V8-powered vehicles, found itself at a crossroads. The company's flagship models, like the Ford Torino and the Ford Maverick, were not initially designed with fuel efficiency in mind. The oil crises threatened to erode Ford's market share as consumers began to seek more economical options.

Strategic Shifts Toward Fuel Efficiency

In response to the crisis, Ford undertook several strategic initiatives to enhance fuel efficiency and meet evolving consumer demands:

- **Introduction of the Ford Pinto (1971):** Although the Pinto was launched before the oil crises, its relevance surged in the wake of rising fuel prices. The Pinto was a compact car designed to compete with imports from Japan and Europe, offering better fuel economy and affordability. Despite its controversial safety issues, the Pinto represented Ford's initial foray into the compact car segment.

- **Development of the Ford Escort (1981):** Recognizing the need for a more modern and fuel-efficient vehicle, Ford developed the Escort, which would later become a staple in the company's lineup. The Escort emphasized lightweight construction, smaller engines, and improved aerodynamics to maximize fuel efficiency.

- **Investment in Fuel-Efficient Technologies:** Ford began investing in technologies such as fuel injection systems, smaller displacement engines, and aerodynamic designs. These innovations aimed to reduce fuel consumption without sacrificing performance.

Challenges Faced by Ford

Despite these efforts, Ford encountered several challenges:

- **Public Perception and Safety Concerns:** The Pinto's safety issues, particularly its susceptibility to fuel tank ruptures in rear-end collisions, tarnished Ford's reputation. The negative publicity surrounding the Pinto's safety compromised consumer trust, making it difficult for Ford to promote its compact models effectively.

- **Economic Pressures:** The high cost of retooling factories and developing new models strained Ford's financial resources. Balancing the need for innovation with profitability was a constant struggle.

- **Competition from Imports:** Japanese automakers like Toyota and Honda were gaining significant traction in the American market with their fuel-efficient and reliable vehicles. Ford had to contend with this stiff competition while trying to establish its own presence in the compact segment.

Chevrolet: Rising to the Challenge

Impact on Chevrolet

Chevrolet, part of the General Motors (GM) conglomerate, faced similar challenges as Ford during the oil crises. Chevrolet's lineup included larger, less fuel-efficient vehicles that were hit hard by the shift in consumer preferences. However, Chevrolet leveraged its resources within GM to implement effective strategies to navigate the crisis.

Strategic Shifts Toward Fuel Efficiency

Chevrolet's response to the oil crises was marked by agility and innovation:

- **Introduction of the Chevrolet Vega (1970):** The Vega was Chevrolet's entry into the compact car market, designed to compete directly with imports. It featured a lightweight design and was available with fuel-efficient inline-four engines. Although the Vega faced quality and reliability issues, it demonstrated Chevrolet's commitment to addressing the demand for economical vehicles.

- **Launch of the Chevrolet Citation (1980):** The Citation was a significant step forward for Chevrolet, offering improved performance, better fuel economy, and enhanced build quality compared to the Vega. It represented a more refined approach to compact car design and aimed to regain consumer confidence.

- **Adoption of Fuel Injection and Smaller Engines:** Chevrolet invested in advanced fuel injection systems and downsized its engine offerings to improve fuel efficiency. These technological advancements were crucial in meeting regulatory standards and consumer expectations.

Challenges Faced by Chevrolet

Chevrolet's journey through the oil crises was not without obstacles:

- **Quality Control Issues:** The Vega's reputation was marred by production problems, including rust, engine failures, and other reliability issues. These defects undermined Chevrolet's efforts to establish credibility in the compact segment.

- **Balancing Performance and Efficiency:** Striking the right balance between performance and fuel efficiency was a delicate task. Chevrolet had to ensure that its vehicles remained appealing to consumers who still desired some level of performance, even as they prioritized economy.

- **Internal GM Struggles:** General Motors faced its own set of internal challenges, including labor disputes, management inefficiencies, and financial constraints. These issues impacted Chevrolet's ability to swiftly implement necessary changes and innovations.

Regulatory Changes and Industry Standards

Introduction of Corporate Average Fuel Economy (CAFE) Standards

In 1975, the U.S. government introduced the Corporate Average Fuel Economy (CAFE) standards as part of the Energy Policy and Conservation Act. These regulations mandated that automakers achieve a certain average fuel efficiency across their fleets, aiming to reduce dependency on oil imports and promote energy conservation.

Impact on Ford and Chevrolet

Both Ford and Chevrolet had to overhaul their manufacturing processes and vehicle designs to comply with CAFE standards:

- **Engine Downsizing and Turbocharging:** Both companies explored smaller engine options and turbocharging to maintain performance levels while improving fuel efficiency. These technologies allowed for reduced displacement without significantly compromising power.

- **Lightweight Materials and Aerodynamic Designs:** Incorporating lightweight materials and enhancing aerodynamics became priorities. Ford and Chevrolet invested in research and development to create vehicles that were both efficient and stylish.

- **Discontinuation of Larger Models:** Some of the larger, less fuel-efficient models were phased out or redesigned to meet the new standards. This shift required significant changes in production lines and supply chains.

Emission Regulations

Alongside fuel economy, emission standards were tightened to address environmental concerns. Both Ford and Chevrolet had to implement cleaner engine technologies to reduce pollutants:

- **Catalytic Converters:** Introduction of catalytic converters became standard, requiring changes in exhaust systems to minimize harmful emissions.

- **Emission Control Technologies:** Both companies invested in advanced emission control systems, including improved fuel injection, exhaust gas recirculation (EGR), and on-board diagnostics (OBD).

Introduction of Compact and Subcompact Cars

Ford's Compact Offerings

Ford's strategy focused on revitalizing its compact car lineup to cater to the new market demands:

- **Ford Pinto (1971):** While the Pinto was criticized for safety flaws, it was Ford's attempt to offer a smaller, more fuel-efficient vehicle. The Pinto featured a compact design and was priced competitively to attract budget-conscious consumers.

- **Ford Maverick (1970):** Introduced as a compact pickup truck, the Maverick aimed to provide the utility of a truck with the fuel efficiency of a car. It offered a variety of engine options, including the inline-six and V8, appealing to a broad range of customers.

Chevrolet's Compact Strategies

Chevrolet also made significant moves to capture the compact car market:

- **Chevrolet Vega (1970):** The Vega was Chevrolet's response to the increasing demand for compact cars. It featured a lightweight design, fuel-efficient engines, and a range of body styles. Despite its initial promise, the Vega struggled with quality issues that hindered its success.

- **Chevrolet Monza (1975):** Replacing the Vega in certain markets, the Monza offered improved styling and performance. It aimed to rectify some of the Vega's shortcomings and provide a more reliable and appealing compact option.

Consumer Behavior and Market Shifts

Changing Priorities

The oil crises fundamentally altered consumer priorities. The focus shifted from vehicle size and performance to fuel economy and practicality. Consumers began to value:

- **Fuel Efficiency:** The ability to travel longer distances on less fuel became a critical factor in purchasing decisions.

- **Affordability:** With rising fuel costs, the overall cost of ownership became a significant consideration.

- **Reliability:** Consumers sought vehicles that were not only economical but also dependable and low-maintenance.

Shift Toward Imports

Japanese and European automakers capitalized on the demand for fuel-efficient and reliable vehicles. Brands like Toyota, Honda, and Volkswagen gained substantial market share by offering cars that met the new consumer demands better than their American counterparts.

Impact on Ford and Chevrolet

The success of imports intensified the competition, forcing Ford and Chevrolet to accelerate their adaptation efforts. It became clear that traditional American automotive strengths—size, power, and durability—were no longer the primary selling points.

Challenges Faced by Ford and Chevrolet

Quality Control and Reliability Issues

Both Ford and Chevrolet struggled with quality control during this period:

- **Ford Pinto Controversy:** The Pinto's safety issues, particularly its propensity for fuel tank ruptures, led to widespread criticism and legal battles. The negative publicity damaged Ford's reputation and highlighted the risks of rushed product development.

- **Chevrolet Vega Defects:** The Vega suffered from rust problems, engine failures, and poor build quality. These issues eroded consumer trust and provided an opening for import manufacturers to gain ground.

Economic Pressures and Production Costs

The need to redesign vehicles for fuel efficiency and comply with regulations increased production costs. Both companies faced:

- **Rising Material Costs:** Lightweight materials and advanced technologies were often more expensive, squeezing profit margins.

- **Retooling and Reengineering:** Significant investments were required to retool factories and redesign vehicle platforms, straining financial resources.

Internal Management and Strategic Decisions

Strategic missteps and internal management issues further compounded the challenges:

- **Ford's Resistance to Change:** Henry Ford II's initial reluctance to embrace smaller, more fuel-efficient models delayed the company's response to the crisis.

- **GM's Bureaucratic Structure:** General Motors' complex and hierarchical management structure slowed decision-making processes, hindering Chevrolet's ability to respond swiftly and effectively.

Recovery and Long-Term Adaptations

Embracing Innovation and Efficiency

Both Ford and Chevrolet realized that the future of the automotive industry lay in innovation and efficiency. They began to implement long-term strategies to ensure sustainability and competitiveness:

- **Investment in Research and Development:** Increased focus on developing fuel-efficient engines, alternative powertrains, and advanced manufacturing techniques.

- **Lean Manufacturing Practices:** Adoption of lean manufacturing principles to reduce waste, improve quality, and enhance production efficiency.

Diversification of Product Lines

To mitigate risks and cater to diverse consumer needs, both companies diversified their product offerings:

- **Introduction of SUVs and Crossovers:** Recognizing the growing demand for versatile and spacious vehicles, Ford and Chevrolet expanded their lineups to include SUVs and crossovers alongside their traditional trucks and cars.

- **Focus on Niche Markets:** Both brands began targeting niche markets, such as performance enthusiasts and eco-conscious consumers, with specialized models and options.

Strengthening Brand Image and Loyalty

Rebuilding consumer trust was paramount. Ford and Chevrolet undertook initiatives to enhance their brand images:

- **Quality Improvement Programs:** Comprehensive quality control measures were implemented to address past deficiencies and ensure higher standards.

- **Customer Engagement and Support:** Enhanced customer service, extended warranties, and loyalty programs aimed to foster stronger relationships with consumers.

Transition to Alternative Fuels and Electric Vehicles

Looking ahead, Ford and Chevrolet began exploring alternative fuels and electric vehicle technologies to stay ahead of regulatory changes and evolving consumer preferences:

- **Hybrid and Electric Prototypes:** Early prototypes and concept cars showcased the companies' commitment to sustainable mobility solutions.

- **Partnerships and Collaborations:** Collaborations with technology firms and research institutions to accelerate the development of electric and hybrid powertrains.

Case Studies: Ford and Chevrolet's Responses

Ford's Strategic Initiatives

Ford Pinto and Its Aftermath

The Ford Pinto serves as a cautionary tale of the dangers of prioritizing speed to market over quality and safety. The Pinto's safety issues led to numerous lawsuits and a damaged reputation. However, Ford learned valuable lessons about the importance of thorough testing and consumer safety, which influenced future product development.

Ford Maverick and Mustang II

The Ford Maverick, introduced in 1970, aimed to blend the utility of a compact car with the versatility of a pickup truck. While it initially struggled with sales, it laid the groundwork for future hybrid models. The Mustang II, launched in 1974, was a direct response to the oil crises, offering improved fuel efficiency and a smaller footprint compared to its predecessor.

Chevrolet's Strategic Initiatives

Chevrolet Vega's Struggles and Lessons

The Vega's quality control issues were a significant setback for Chevrolet. However, the lessons learned from the Vega's failures informed the development of the Chevrolet Citation, which featured better build quality and more reliable performance. The Citation aimed to restore consumer confidence and compete more effectively with both domestic and import models.

Chevrolet Citation and Camaro Adjustments

The Citation represented Chevrolet's commitment to

improving its compact offerings. By addressing the Vega's shortcomings, Chevrolet was able to offer a more competitive and reliable vehicle. Meanwhile, the Camaro continued to evolve, incorporating fuel-efficient engines and aerodynamic designs to maintain its appeal amid changing market conditions.

Regulatory Compliance and Environmental Responsibility

Meeting CAFE Standards

Both Ford and Chevrolet had to navigate the complexities of meeting Corporate Average Fuel Economy (CAFE) standards. This involved:

- **Fleet Fuel Efficiency Management:** Balancing the fuel efficiency of passenger cars with the less efficient trucks to meet overall fleet targets.

- **Strategic Model Planning:** Phasing out high-displacement engines and introducing more efficient powertrains across their vehicle lineups.

Emission Control Technologies

To comply with stringent emission regulations, both companies invested in advanced emission control technologies:

- **Catalytic Converters:** Standardizing catalytic converters across models to reduce harmful emissions.

- **On-Board Diagnostics (OBD):** Developing OBD systems to monitor and control engine performance, ensuring compliance with emission standards.

Corporate Environmental Initiatives

Beyond regulatory compliance, Ford and Chevrolet began to incorporate environmental responsibility into their corporate strategies:

- **Sustainable Manufacturing Practices:** Implementing energy-efficient processes, reducing waste, and utilizing recyclable materials in vehicle production.

- **Research into Alternative Fuels:** Exploring biofuels, hydrogen, and electric powertrains as potential alternatives to traditional gasoline engines.

Economic and Competitive Implications

Market Share Dynamics

The oil crises and subsequent market shifts had significant implications for market share:

- **Ford's Resilience:** Despite setbacks like the Pinto controversy, Ford's early investment in compact cars and fuel-efficient technologies allowed it to maintain a strong presence in the market.

- **Chevrolet's Recovery:** Chevrolet's ability to learn from the Vega's failures and improve its compact offerings helped it regain market share and compete effectively against both Ford and import brands.

Financial Performance

The economic pressures of the 1970s impacted the financial performance of both companies:

- **Revenue Declines:** Reduced demand for large, fuel-inefficient vehicles led to declines in revenue from key segments.

- **Cost Management:** Both companies had to implement cost-cutting measures, streamline operations, and optimize supply chains to remain profitable.

Competitive Strategies

To stay competitive, Ford and Chevrolet adopted various strategies:

- **Innovation and Differentiation:** Developing unique features, enhancing vehicle performance, and offering extensive customization options to differentiate their products.

- **Strategic Marketing:** Shifting marketing messages to emphasize fuel efficiency, reliability, and affordability, aligning with consumer priorities.

- **Collaborations and Partnerships:** Engaging in partnerships with technology firms and research institutions to accelerate innovation and develop advanced vehicle technologies.

Long-Term Impact and Legacy

Transformation of the Automotive Landscape

The oil crises of the 1970s were a turning point for the automotive industry, leading to lasting changes in vehicle design, manufacturing practices, and consumer expectations. Ford and Chevrolet's responses to these challenges played a crucial role in shaping the modern automotive landscape.

Shift Toward Compact and Efficient Vehicles

The emphasis on fuel efficiency and compact design laid the foundation for the future success of smaller vehicles. Both Ford and Chevrolet continued to invest in compact and subcompact cars, recognizing their importance in a market increasingly focused on economy and sustainability.

Foundation for Future Innovations

The lessons learned during the oil crises informed future innovations, including the development of hybrid and electric vehicles. Ford and Chevrolet's early efforts in improving fuel efficiency and reducing emissions positioned them to adapt more readily to subsequent environmental and technological trends.

Cultural and Environmental Responsibility

The 1970s oil crises heightened awareness of environmental issues and the need for sustainable practices. Ford and Chevrolet's initiatives toward environmental responsibility helped establish a corporate culture that values sustainability, influencing their long-term strategies and public image.

Conclusion: Navigating Through Crisis with Resilience and Innovation

The 1970s oil crises were a defining moment for the automotive industry, challenging established norms and pushing manufacturers to innovate in unprecedented ways. Ford and Chevrolet, two of America's automotive giants, faced these challenges head-on, adapting their strategies to meet the new demands of fuel efficiency and environmental responsibility.

Ford's journey through the crises was marked by both

missteps and strategic pivots, illustrating the complexities of adapting to rapidly changing market conditions. Chevrolet's response, though initially fraught with quality control issues, ultimately demonstrated resilience and a commitment to learning from past mistakes.

The legacy of this era is evident in the continued focus on fuel efficiency, the diversification of vehicle lineups, and the ongoing pursuit of sustainable technologies. The ability of Ford and Chevrolet to navigate the oil crises not only ensured their survival but also set the stage for future innovations that would keep them at the forefront of the automotive industry.

As we move forward in our exploration of the Ford versus Chevrolet rivalry, the lessons from the 1970s oil crises highlight the importance of adaptability, innovation, and strategic foresight. These principles would continue to guide both companies as they faced new challenges and seized emerging opportunities in the decades to come, ensuring that the rivalry remained as dynamic and influential as ever.

Chapter 11: The Evolution of Design and Technology in the '80s and '90s

As the United States approached the end of the 20th century, the automotive landscape was undergoing a profound transformation. The 1980s and 1990s were decades of significant advancements in technology, safety, and design philosophies. Amidst shifting consumer preferences, stringent environmental regulations, and fierce competition, Ford and Chevrolet embarked on a journey of reinvention to maintain their dominance in the market. This chapter explores the pivotal developments in automotive technology and design during these decades and examines how Ford and Chevrolet strategized to stay relevant and competitive.

Technological Advancements: Driving Innovation

Engine Efficiency and Performance

The '80s and '90s marked a period of relentless pursuit for engine efficiency without compromising performance. Both Ford and Chevrolet invested heavily in research and development to create engines that delivered better fuel economy and lower emissions, responding to rising fuel prices and environmental concerns.

Ford's EcoBoost Initiative

While the term "EcoBoost" would not be coined until the 2000s, Ford's efforts in the '80s and '90s laid the groundwork for turbocharged, direct-injection engines. Innovations such as variable valve timing and electronic fuel injection improved engine responsiveness and efficiency.

- **Ford Taurus SHO (1989)**: Introduced with a twin-turbocharged V6 engine, the SHO (Super High Output) variant of the Taurus showcased Ford's commitment to blending performance with efficiency.

This model became a favorite among enthusiasts for its balance of power and practicality.

Chevrolet's Small-Block Evolution

Chevrolet continued to refine its legendary small-block V8 engines, focusing on enhancing power output while reducing fuel consumption.

- **Chevrolet LS Series (1997)**: The introduction of the LS series engines revolutionized Chevrolet's performance offerings. These engines featured aluminum blocks, improved airflow designs, and advanced fuel injection systems, providing superior performance and efficiency.

Electronic Advancements and Computerization

The integration of electronics into automotive systems became a hallmark of the '80s and '90s, enhancing vehicle performance, safety, and user experience.

Ford's On-Board Diagnostics (OBD)

Ford was among the first to adopt comprehensive OBD systems, which allowed for real-time monitoring of engine performance and emissions.

- **Ford Taurus (1996)**: Equipped with OBD-II, the Taurus featured advanced diagnostics that not only improved maintenance but also aided in reducing emissions, aligning with regulatory requirements.

Chevrolet's Electronic Stability Control

Chevrolet focused on enhancing vehicle stability and safety through electronic systems.

- **Chevrolet Silverado (1999)**: The Silverado introduced electronic stability control (ESC) as an optional feature, providing drivers with enhanced control during adverse conditions and reducing the likelihood of accidents.

Fuel Injection and Emission Control Technologies

Both manufacturers prioritized advancements in fuel injection and emission control to meet tightening environmental standards.

Direct Fuel Injection

- **Ford's Direct Injection Engines**: Ford began experimenting with direct fuel injection systems, which allowed for more precise fuel delivery, improving combustion efficiency and reducing emissions.

- **Chevrolet's LE1 and LT Series**: Chevrolet introduced the LE1 and later the LT series engines with multi-point fuel injection, enhancing performance and fuel economy.

Catalytic Converters and Emission Reductions

Both companies invested in catalytic converter technology to minimize harmful emissions.

- **Ford's Catalytic Converters**: Implemented across their vehicle lineup, these converters significantly reduced pollutants, helping Ford comply with the Clean Air Act.

- **Chevrolet's Emission Control Systems**: Chevrolet introduced advanced emission control systems, including oxygen sensors and exhaust gas recirculation (EGR), to further decrease emissions.

Safety Innovations: Protecting Lives

The '80s and '90s saw significant advancements in automotive safety, driven by consumer demand and regulatory mandates. Ford and Chevrolet were at the forefront of integrating these innovations into their vehicle lineups.

Airbag Systems

Airbags became a standard safety feature during this period, providing critical protection in the event of a collision.

Ford's Airbag Implementation

- **Ford Escort (1988)**: Ford began equipping the Escort with dual front airbags, setting a precedent for passenger protection in compact cars.

- **Ford Explorer (1991)**: As SUVs gained popularity, the Explorer incorporated airbags alongside other safety features, catering to families seeking both utility and safety.

Chevrolet's Airbag Adoption

- **Chevrolet Cavalier (1987)**: The Cavalier was among the first Chevrolet models to offer airbags as an option, enhancing safety without significantly increasing costs.

- **Chevrolet Suburban (1992)**: The Suburban, a staple in Chevrolet's lineup, introduced airbags along with reinforced frames and advanced restraint systems.

Anti-lock Braking Systems (ABS)

ABS technology improved vehicle control during emergency braking by preventing wheel lock-up.

Ford's ABS Integration

- **Ford Taurus (1992)**: Ford introduced ABS as an option on the Taurus, providing enhanced braking performance and safety.

- **Ford F-150 (1997)**: The introduction of ABS on the F-150 improved its handling and safety, particularly during towing and off-road operations.

Chevrolet's ABS Implementation

- **Chevrolet C/K Series (1980s)**: Chevrolet began offering ABS on its C/K series trucks, enhancing their safety credentials for commercial and personal use.

- **Chevrolet Lumina (1990)**: The Lumina sedan featured ABS as standard equipment, aligning with Ford's offerings and meeting consumer expectations.

Crashworthiness and Structural Integrity

Advancements in vehicle structure and materials improved crashworthiness, ensuring better protection for occupants.

Ford's Safety Enhancements

- **Ford Crown Victoria (1992)**: Designed with a robust body-on-frame construction, the Crown Victoria offered superior crash protection, making it a popular choice for law enforcement and taxi services.

- **Ford Explorer (1991)**: The Explorer's design

incorporated crumple zones and reinforced passenger compartments, enhancing overall vehicle safety.

Chevrolet's Structural Innovations

- **Chevrolet Silverado (1999)**: The Silverado featured a reinforced frame and advanced side-impact protection, ensuring occupant safety during collisions.

- **Chevrolet Lumina (1990)**: The Lumina employed high-strength steel and improved structural designs to enhance crash performance.

Design Philosophies: Aesthetic and Functional Evolution

The '80s and '90s were a time of evolving design philosophies, with a focus on blending aesthetics with functionality. Ford and Chevrolet adapted their design strategies to meet changing consumer tastes and market demands.

Aerodynamics and Fuel Efficiency

Improved aerodynamics contributed to better fuel efficiency and performance, becoming a key design consideration.

Ford's Aerodynamic Designs

- **Ford Taurus (1986)**: Known for its aerodynamic shape, the Taurus reduced drag and improved fuel economy, setting a trend for future sedan designs.

- **Ford Escort (1990)**: The Escort's streamlined design enhanced its fuel efficiency and market appeal in the compact segment.

Chevrolet's Streamlined Aesthetics

- **Chevrolet Corsica (1987)**: The Corsica featured sleek lines and aerodynamic profiles, improving its performance and appeal in the midsize market.

- **Chevrolet Camaro (1993)**: The fourth-generation Camaro adopted more aerodynamic contours, balancing aggressive styling with functional efficiency.

Ergonomics and Interior Comfort

Enhanced ergonomics and interior comfort became priorities, reflecting a shift towards consumer-centric designs.

Ford's Interior Innovations

- **Ford Windstar (1995)**: As minivans gained popularity, the Windstar introduced a spacious, flexible interior with sliding doors and user-friendly controls.

- **Ford Explorer (1991)**: The Explorer's interior was designed for comfort and practicality, featuring adjustable seating and advanced infotainment systems.

Chevrolet's Interior Enhancements

- **Chevrolet Lumina APV (1989)**: The Lumina APV showcased innovative interior designs with a versatile seating arrangement and modern controls.

- **Chevrolet Silverado (1999)**: The Silverado offered a refined interior with ergonomic controls, premium materials, and advanced infotainment options.

Styling Trends: From Boxy to Sleek

Automotive styling transitioned from the boxy designs of the '70s to more sleek and modern aesthetics, influenced by consumer preferences and competitive pressures.

Ford's Sleek Lineage

- **Ford Taurus (1986)**: The Taurus featured a groundbreaking aerodynamic design with smooth lines and a rounded silhouette, earning accolades for its styling.

- **Ford Explorer (1991)**: The Explorer's sleek yet rugged design appealed to a wide range of consumers, blending SUV utility with modern aesthetics.

Chevrolet's Modern Appeal

- **Chevrolet Corsica (1987)**: The Corsica's modern design emphasized aerodynamics and style, aligning with contemporary trends.

- **Chevrolet Silverado (1999)**: The Silverado's design combined toughness with sophistication, appealing to both commercial users and personal buyers seeking a stylish truck.

Brand Strategies: Reinventing for Relevance

To navigate the technological and design shifts of the '80s and '90s, Ford and Chevrolet implemented strategic initiatives aimed at reinventing their brands and staying relevant in a competitive market.

Ford's Strategic Initiatives

Diversification of Product Lines

Ford expanded its product portfolio to cater to diverse market segments, ensuring that it met the evolving needs of consumers.

- **Introduction of the Ford Taurus**: The Taurus became a flagship model, representing Ford's commitment to innovation and quality in the midsize sedan market.

- **Expansion into SUVs and Minivans**: The launch of the Explorer and Windstar allowed Ford to tap into the growing demand for versatile and family-friendly vehicles.

Focus on Quality and Reliability

In response to past quality issues, Ford prioritized enhancing the quality and reliability of its vehicles.

- **Ford Quality Improvement Programs**: Implemented comprehensive quality control measures and invested in better manufacturing processes to reduce defects and improve customer satisfaction.

- **Customer Feedback Integration**: Actively sought and incorporated customer feedback into vehicle design and production, fostering a customer-centric approach.

Marketing and Branding Overhaul

Ford revamped its marketing strategies to better align with contemporary consumer values and preferences.

- **Emphasis on Innovation**: Highlighted technological advancements and design innovations in advertising campaigns.

- **Building Emotional Connections**: Focused on creating emotional connections with consumers through storytelling and highlighting real-life applications of Ford vehicles.

Chevrolet's Strategic Initiatives

Revamping the Product Lineup

Chevrolet undertook significant efforts to refresh and modernize its vehicle lineup, addressing both performance and efficiency.

- **Chevrolet Corvette (C4 and C5 Generations)**: Redesigned the Corvette with improved aerodynamics, advanced materials, and enhanced performance, maintaining its status as a premier sports car.

- **Introduction of the Chevrolet Silverado**: Reinvented the Silverado with modern design, advanced technology, and superior performance, solidifying its position in the truck market.

Embracing Technology and Innovation

Chevrolet leveraged technological advancements to enhance vehicle performance, safety, and user experience.

- **Advanced Infotainment Systems**: Integrated state-of-the-art infotainment and connectivity features, catering to tech-savvy consumers.

- **Emission and Fuel Efficiency Technologies**:

Invested in cleaner engine technologies and emission control systems to comply with regulations and meet consumer demands for eco-friendly vehicles.

Strengthening Brand Loyalty and Community Engagement

Chevrolet focused on building strong relationships with its customers and fostering brand loyalty through community engagement and exceptional service.

- **Chevrolet Owner Clubs and Events**: Organized events, rallies, and meet-ups to strengthen the community among Chevrolet owners and enthusiasts.

- **Customer Service Enhancements**: Improved aftersales services, including warranties, maintenance packages, and customer support initiatives, to enhance overall ownership experience.

Key Models Illustrating Evolution

Ford Taurus (1986–2007)

The Ford Taurus was a pivotal model in Ford's lineup, embodying the company's shift towards modern design, advanced technology, and enhanced safety.

- **Design Innovation**: Pioneered aerodynamic styling with a rounded silhouette, setting a new standard for mid-size sedans.

- **Technological Features**: Introduced electronic systems like automatic climate control, advanced suspension systems, and later, hybrid powertrains.

- **Market Impact**: The Taurus became one of the best-selling cars in America, influencing the design and marketing strategies of other automakers.

Chevrolet Corvette (C4 and C5 Generations)

Chevrolet's Corvette continued to evolve, maintaining its reputation as America's sports car while integrating cutting-edge technology and design.

- **C4 Generation (1984–1996)**: Featured a more angular design, improved aerodynamics, and advanced electronics. Introduced the LS1 engine, which significantly boosted performance.

- **C5 Generation (1997–2004)**: Focused on structural rigidity and weight reduction, the C5 Corvette introduced the LS2 engine and advanced suspension systems, enhancing both performance and handling.

Ford Explorer (1991–Present)

The Ford Explorer played a crucial role in popularizing the SUV segment, blending utility with comfort and modern design.

- **First Generation (1991–1994)**: Offered a versatile interior, robust performance options, and rugged styling, making it a favorite among families and outdoor enthusiasts.

- **Subsequent Generations**: Continuously evolved with improved technology, safety features, and design refinements to meet changing consumer demands.

Chevrolet Silverado (1998–Present)

The Chevrolet Silverado redefined the truck market with its blend of performance, comfort, and advanced technology.

- **First Generation (1999–2006)**: Emphasized improved ride quality, towing capabilities, and modern styling.

- **Second Generation (2007–2014)**: Introduced more refined aesthetics, enhanced performance options, and advanced safety features.

- **Third Generation (2015–Present)**: Integrated state-of-the-art technology, fuel-efficient engines, and versatile configurations to cater to a wide range of consumers.

Challenges Faced During Reinvention

Balancing Tradition and Innovation

Both Ford and Chevrolet faced the challenge of maintaining their legacy while embracing new technologies and design philosophies.

- **Ford's Legacy Models**: Balancing the innovative Taurus and Explorer with traditional models like the F-Series required careful strategy to avoid alienating loyal customers.

- **Chevrolet's Corvette Heritage**: Preserving the Corvette's iconic status while integrating modern technologies demanded a delicate balance between tradition and progress.

Economic Fluctuations and Market Pressures

The '80s and '90s were marked by economic volatility, including recessions and changing consumer spending patterns.

- **Ford's Financial Strategies**: Implemented cost-cutting measures and streamlined operations to maintain profitability during economic downturns.

- **Chevrolet's Market Positioning**: Focused on leveraging its diverse product lineup to mitigate risks associated with market fluctuations.

Technological Integration and Production Challenges

Integrating advanced technologies into vehicles posed production challenges and required significant investments.

- **Ford's Manufacturing Adaptations**: Upgraded manufacturing facilities to accommodate new technologies and improve production efficiency.

- **Chevrolet's Quality Control**: Addressed quality issues from previous decades by enhancing manufacturing processes and implementing stricter quality control measures.

Brand Reinvention: Staying Relevant in a Changing Market

Ford's Rebranding Efforts

Emphasis on Quality and Reliability

In response to past quality issues, Ford undertook initiatives to enhance the quality and reliability of its vehicles.

- **Ford Quality Assurance Programs**: Implemented comprehensive quality checks and employee training programs to reduce defects and improve manufacturing standards.

- **Customer Satisfaction Focus**: Prioritized customer feedback and satisfaction, integrating it into product development and service offerings.

Adoption of Modern Design Philosophies

Ford embraced contemporary design trends, focusing on aerodynamics, ergonomics, and aesthetics.

- **Collaborations with Renowned Designers**: Partnered with leading automotive designers to create visually appealing and functional vehicle designs.

- **Interior Comfort Enhancements**: Invested in improving interior materials, layouts, and technological features to enhance the driving experience.

Chevrolet's Rebranding Efforts

Innovation and Technological Leadership

Chevrolet positioned itself as a leader in automotive innovation, integrating cutting-edge technologies into its vehicles.

- **Advanced Infotainment Systems**: Introduced user-friendly infotainment systems with features like touchscreen interfaces, smartphone integration, and premium sound systems.

- **Fuel-Efficient and Eco-Friendly Models**: Developed vehicles with improved fuel economy and lower emissions, aligning with environmental trends.

Revitalizing the Brand Image

Chevrolet focused on revitalizing its brand image to appeal to a broader and more diverse consumer base.

- **Marketing Campaigns Highlighting Versatility**: Emphasized the versatility of Chevrolet vehicles, showcasing their suitability for both work and leisure.

- **Engagement with Younger Audiences**: Leveraged sponsorships, events, and digital marketing to connect with younger consumers and build long-term brand loyalty.

Case Studies: Ford and Chevrolet's Key Models in the '80s and '90s

Ford Taurus: A Revolution in Design and Technology

- **Introduction (1986)**: The Taurus introduced a new design language with its aerodynamic shape, moving away from the boxy designs of the '70s. It set new standards for mid-size sedans with its spacious interior and advanced features.

- **Technological Innovations**: Featured electronic controls, improved suspension systems, and safety enhancements like airbags and anti-lock brakes.

- **Market Impact**: Became one of the best-selling cars in America, influencing other automakers to adopt similar design and technological trends.

Chevrolet Corvette C4 and C5 Generations

- **C4 Corvette (1984–1996)**: Introduced a more aerodynamic and modern design, incorporating advanced materials and technologies like digital instrument clusters.

- **C5 Corvette (1997–2004)**: Focused on improving performance and handling with a new LS1 engine, independent rear suspension, and enhanced aerodynamics.

- **Legacy**: Both generations maintained the Corvette's status as America's premier sports car, blending performance with cutting-edge technology.

Ford F-Series and Chevrolet Silverado

- **Ford F-Series Evolution**: Continued to dominate the truck market with improvements in engine performance, towing capacity, and interior comfort. The introduction of the F-150 in 1975 set a benchmark for versatility and reliability.

- **Chevrolet Silverado Evolution**: Reinvented the Silverado with modern design elements, advanced technology, and superior performance options. The Silverado's focus on comfort and capability made it a strong competitor to the F-Series.

Chevrolet Cavalier and Ford Escort

- **Chevrolet Cavalier (1982–2005)**: A compact car designed to compete with imports, featuring fuel-efficient engines and practical design. It became a popular choice for budget-conscious consumers.

- **Ford Escort (1981–2000)**: Ford's compact offering that emphasized fuel efficiency and practicality, evolving through multiple generations to incorporate modern features and styling.

Impact on the Automotive Industry and Consumer Expectations

Setting New Standards

The advancements made by Ford and Chevrolet during the '80s and '90s set new benchmarks for the entire automotive industry.

- **Fuel Efficiency and Emissions**: Their efforts to improve fuel economy and reduce emissions influenced regulatory standards and encouraged other automakers to prioritize environmental responsibility.

- **Safety Innovations**: Introduction of airbags, ABS, and other safety features became standard across the industry, enhancing overall vehicle safety.

Influencing Consumer Expectations

Consumers began to expect more from their vehicles, including better fuel economy, advanced technology, enhanced safety, and stylish design.

- **Technological Integration**: Features like electronic fuel injection, computerized engine management, and advanced infotainment systems became sought-after amenities.

- **Design Aesthetics**: Aerodynamic and ergonomic designs set new trends, making vehicles not only functional but also visually appealing.

Competitive Landscape Transformation

The strategies and innovations of Ford and Chevrolet reshaped the competitive landscape, pushing other automakers to innovate and improve their offerings.

- **Rise of Imports**: Japanese and European manufacturers responded to Ford and Chevrolet's innovations by introducing their own advanced and fuel-efficient models.

- **Global Influence**: Ford and Chevrolet's advancements had a global impact, influencing automotive design and technology standards worldwide.

Conclusion: Reinvention and Resilience

The '80s and '90s were transformative decades for the automotive industry, marked by significant technological advancements, enhanced safety features, and evolving design philosophies. Ford and Chevrolet navigated these changes with strategic initiatives aimed at reinvention and adaptation. By embracing innovation, prioritizing quality and safety, and responding to shifting consumer demands, both companies reinforced their positions as industry leaders.

The legacy of this era is evident in the continued success and evolution of key models like the Ford Taurus, Chevrolet Corvette, Ford F-Series, and Chevrolet Silverado. These vehicles not only defined their respective brands but also set enduring standards for the automotive industry as a whole.

As we move forward in exploring the Ford versus Chevrolet rivalry, the lessons from the '80s and '90s highlight the importance of adaptability, innovation, and customer-centric strategies. The ability of Ford and Chevrolet to reinvent themselves during challenging times ensured their resilience and continued relevance, paving the way for future chapters that will delve into the turn of the millennium and beyond.

Chapter 12: The Revival of the Classics

As the turn of the millennium approached, the automotive landscape was witnessing a powerful resurgence of nostalgia. Consumers were increasingly drawn to the iconic models of the past, seeking to recapture the spirit and excitement that defined earlier eras. This wave of nostalgia was particularly evident in the resurgence of classic muscle cars, with Ford and Chevrolet at the forefront of revitalizing their legendary models—the Mustang and Camaro. This chapter explores the late 1990s and early 2000s resurgence of interest in these classics, examining how both brands modernized their offerings to appeal to new generations while honoring their storied legacies.

A Nostalgic Resurgence: The Late '90s and Early 2000s

Cultural Shift Toward Nostalgia

The late 1990s and early 2000s were characterized by a cultural shift toward nostalgia, as consumers yearned for the simplicity and excitement of earlier decades. This sentiment was fueled by various factors:

- **Media Influence**: Movies, television shows, and music from the '60s and '70s enjoyed renewed popularity, rekindling interest in the era's automotive icons.

- **Generational Transition**: Baby boomers and Generation X, who had grown up with classic muscle cars, began influencing the market with their purchasing power and desire to relive their youth.

- **Automotive Heritage Appreciation**: There was a growing appreciation for automotive heritage, with enthusiasts and collectors seeking to preserve and celebrate classic models.

Market Opportunity for Classic Revivals

Recognizing this burgeoning interest, Ford and Chevrolet saw an opportunity to reinvigorate their classic models. By leveraging their rich histories and iconic designs, both brands aimed to create modern interpretations that would resonate with contemporary consumers while paying homage to their predecessors.

Modern Interpretations: Updating the Mustang and Camaro

Ford Mustang: Balancing Legacy and Innovation

Fifth Generation Mustang (1994–2004)

The fifth-generation Ford Mustang, known internally as the SN-95, marked a significant evolution from its predecessors. Launched in 1994, this generation sought to blend the Mustang's classic design elements with modern engineering and technology.

- **Design Enhancements**: The SN-95 retained the Mustang's signature long hood and short deck but featured more aerodynamic lines and refined styling cues. Improvements included updated headlights, taillights, and a redesigned interior that offered greater comfort and functionality.

- **Engine and Performance Upgrades**: Engine options were expanded to include more powerful V6 and V8 engines. The introduction of the Cobra model in 1996, equipped with a 4.6-liter V8 engine producing 305 horsepower, underscored Ford's commitment to performance.

Special Editions and Variants

To capitalize on the nostalgia wave, Ford introduced several special editions and performance variants of the Mustang during this period:

- **SVT Cobra (1999–2004)**: Developed by Ford's Special Vehicle Team (SVT), the Cobra featured a high-performance 5.4-liter V8 engine producing up to 385 horsepower. Enhanced suspension, brakes, and aerodynamics made the Cobra a formidable competitor on both the street and the track.

- **Bullitt (2001–2004)**: Inspired by the classic 1968 Mustang GT driven by Steve McQueen in the film "Bullitt," this limited-edition model featured unique styling elements, including a hood bulge, matte-black paint options, and a tuned 4.6-liter V8 engine delivering 300 horsepower. The Bullitt Mustang became an instant classic, revered by enthusiasts for its connection to automotive history.

- **Mach 1 (2003–2004)**: Revived as a homage to the original Mach 1 from the 1960s, this variant emphasized performance with a 4.6-liter V8 engine, enhanced suspension, and aggressive styling cues. The Mach 1 bridged the gap between the standard Mustang and the high-performance Cobra, appealing to a broader audience.

Technological Integrations

The SN-95 Mustang incorporated several technological advancements to enhance performance, safety, and user experience:

- **Advanced Suspension Systems**: Improved suspension geometry and components provided better handling and ride quality, making the Mustang more agile and responsive.

- **Safety Features**: Introduction of side-impact airbags, anti-lock brakes (ABS), and traction control systems increased the vehicle's safety profile, aligning with evolving regulatory standards.

- **Infotainment and Connectivity**: Upgraded audio systems, optional navigation, and improved ergonomics enhanced the overall driving experience, making the Mustang more appealing to modern consumers.

Chevrolet Camaro: Reinforcing the Legacy

Fourth Generation Camaro (1993–2002)

Chevrolet's fourth-generation Camaro, launched in 1993, played a crucial role in revitalizing the model after a hiatus following the discontinuation of the fifth-generation Camaro in 2002. This generation focused on updating the Camaro's design and performance to compete effectively with the Mustang.

- **Design Overhaul**: The fourth-generation Camaro featured more refined and aerodynamic styling compared to its predecessor. Key updates included new front and rear fascias, updated lighting elements, and a more streamlined silhouette that maintained the Camaro's aggressive stance.

- **Engine Options and Performance**: Engine offerings were diversified to include fuel-efficient four-cylinder engines alongside potent V6 and V8 options. The introduction of the Z/28 variant in 1996, equipped with a 5.7-liter LS1 V8 engine producing 275 horsepower, underscored Chevrolet's commitment to performance.

Special Editions and Variants

Chevrolet introduced several special editions and performance-focused variants to appeal to enthusiasts and leverage the Camaro's classic appeal:

- **Z/28 (1996–2002)**: Designed for racing homologation, the Z/28 featured a high-revving 5.7-liter LS1 V8 engine, lightweight components, and enhanced suspension systems. Its success in the Trans-Am Series bolstered the Camaro's performance credentials.

- **LS1 (1998–2002)**: The LS1 Camaro offered a 5.7-liter V8 engine producing up to 300 horsepower. Enhanced performance features, including improved brakes and suspension, made the LS1 a favorite among performance enthusiasts.

- **RS and SS Packages**: The Rally Sport (RS) and Super Sport (SS) packages provided aesthetic and performance enhancements, such as upgraded wheels, spoilers, and engine tuning, allowing customers to personalize their Camaros to their preferences.

Technological Integrations

The fourth-generation Camaro incorporated several technological advancements to enhance its appeal and performance:

- **Advanced Engine Technology**: Introduction of the LS1 engine featured improved fuel injection, better airflow designs, and enhanced cooling systems, providing a significant boost in performance and efficiency.

- **Enhanced Safety Features**: Implementation of side-impact airbags, ABS, and reinforced structures improved the Camaro's safety profile, making it more appealing to a broader audience.

- **Modern Infotainment Systems**: Upgraded audio systems, optional navigation, and improved interior ergonomics enhanced the driving experience, aligning the Camaro with contemporary consumer expectations.

Marketing Strategies: Tapping into Nostalgia and Innovation

Ford's Marketing Approach

Ford's marketing strategies during the late '90s and early 2000s focused on blending nostalgia with modern innovation:

- **Heritage Celebrations**: Campaigns highlighted the Mustang's storied past, celebrating its legacy while showcasing new advancements. Advertisements often featured classic Mustang imagery alongside modern models, emphasizing continuity and evolution.

- **Performance Emphasis**: Marketing efforts for special editions like the SVT Cobra and Bullitt focused on performance capabilities, leveraging motorsports success and celebrity endorsements to enhance desirability.

- **Youth-Centric Campaigns**: Targeting younger consumers through sponsorships of sports events, music festivals, and media placements in popular culture helped Ford maintain the Mustang's appeal across generations.

Chevrolet's Marketing Approach

Chevrolet's marketing strategies mirrored Ford's focus on heritage and performance while emphasizing innovation and versatility:

- **Legacy and Performance**: Advertisements for the Camaro highlighted its racing heritage and performance credentials, leveraging success in motorsports to build brand credibility.

- **Special Editions Promotion**: Campaigns for models like the Z/28 and LS1 emphasized exclusivity and performance, appealing to enthusiasts seeking high-performance vehicles with a unique edge.

- **Modern Design Focus**: Marketing materials showcased the Camaro's modern design updates, highlighting its blend of classic muscle car aesthetics with contemporary styling and technology.

Public Reception and Sales Impact

Ford Mustang's Continued Success

The revitalized Mustang models enjoyed robust sales and positive reception:

- **Sales Figures**: The introduction of special editions and performance variants drove strong sales, with the Mustang maintaining its position as a top-selling sports car in the United States.

- **Critical Acclaim**: Reviews praised the Mustang's balance of classic styling and modern performance, commending models like the SVT Cobra and Bullitt for their engineering excellence and heritage connections.

- **Cultural Impact**: The Mustang continued to be a cultural icon, featured prominently in movies, television shows, and music, reinforcing its status as a symbol of American automotive prowess.

Chevrolet Camaro's Resurgence

The fourth-generation Camaro also saw significant success, though it faced stiff competition from the Mustang:

- **Sales Figures**: While the Camaro did not surpass the Mustang in sales, it held a strong market presence, particularly with performance-oriented models like the Z/28 and LS1 driving interest.

- **Enthusiast Reception**: Camaro enthusiasts lauded the model for its improved performance, handling, and design, appreciating the balance between modern enhancements and classic muscle car attributes.

- **Motorsports Success**: Success in racing series like Trans-Am and drag racing enhanced the Camaro's reputation, translating into increased consumer interest and sales.

Legacy Models: Icons of Performance and Heritage

Ford Mustang Shelby GT350 and GT500

The Shelby variants of the Mustang continued to embody high performance and exclusivity:

- **GT350**: Featuring a meticulously tuned 5.4-liter V8 engine, the GT350 offered exceptional performance and handling. Limited production runs and racing success made it a coveted model among enthusiasts.

- **GT500**: Introduced with a 5.8-liter V8 engine producing over 500 horsepower, the GT500 represented the pinnacle of Mustang performance. Advanced technologies like dual overhead camshafts and high-performance suspension systems solidified its status as a true muscle car.

Chevrolet Camaro Z/28 and LS1

Chevrolet's special editions maintained the Camaro's competitive edge:

- **Z/28**: Designed for racing homologation, the Z/28's performance-oriented features and lightweight construction made it a favorite in motorsports, enhancing its street credibility.

- **LS1**: The LS1 Camaro offered a significant power boost with its 5.7-liter V8 engine, coupled with performance enhancements that appealed to both street enthusiasts and track drivers.

Technological and Design Legacy

Influence on Future Models

The innovations and design philosophies adopted during the late '90s and early 2000s laid the groundwork for future developments in both the Mustang and Camaro lineups:

- **Ford Mustang**: The balance of heritage and modernity established during this period influenced subsequent generations, leading to models that continued to honor the Mustang's legacy while integrating advanced technologies and design elements.

- **Chevrolet Camaro**: The emphasis on performance

and racing heritage set the stage for future Camaro generations, ensuring that the model remained relevant and competitive in an evolving market.

Consumer Expectations and Industry Standards

The revival of classic models elevated consumer expectations, setting new standards for performance, design, and technology:

- **Performance Benchmarks**: The high-performance variants set benchmarks for speed, handling, and power, pushing other automakers to enhance their offerings to remain competitive.

- **Design Trends**: The integration of aerodynamic and ergonomic design elements influenced broader automotive design trends, promoting a blend of aesthetics and functionality.

- **Technological Integration**: Features introduced during this era, such as advanced fuel injection systems, improved safety features, and modern infotainment systems, became standard expectations across the industry.

Cultural Significance and Enduring Popularity

The revitalized Mustang and Camaro models not only enjoyed commercial success but also became enduring symbols of American automotive culture:

- **Media Representation**: Continued appearances in films, television, and music reinforced the cultural significance of both models, embedding them deeper into the public consciousness.

- **Collector Appeal**: Special editions and performance

variants became highly sought after by collectors, ensuring the preservation of these models as automotive legends.

- **Community and Enthusiast Engagement**: Strong communities and enthusiast groups formed around the Mustang and Camaro, fostering a sense of camaraderie and shared passion that contributed to their enduring popularity.

Conclusion: A Harmonious Blend of Past and Present

The late 1990s and early 2000s were pivotal years for the Ford Mustang and Chevrolet Camaro, as both models experienced a renaissance that blended classic appeal with modern innovation. By honoring their rich histories while embracing technological advancements and contemporary design philosophies, Ford and Chevrolet successfully revitalized their iconic models, ensuring their relevance and desirability for new generations of consumers.

This revival not only strengthened the brands' market positions but also enriched the broader automotive culture, celebrating the enduring legacy of classic muscle cars while paving the way for future innovations. As we continue to explore the Ford versus Chevrolet rivalry, the lessons from this era underscore the importance of adaptability, heritage preservation, and continuous innovation in maintaining automotive excellence.

The Mustang and Camaro remain shining examples of how automakers can honor their past while driving forward into the future, embodying the spirit of competition and passion that has defined the Ford and Chevrolet rivalry for decades.

Chapter 13: Globalization and International Markets

As the automotive industry evolved through the latter part of the 20th century, globalization emerged as a defining trend reshaping how manufacturers operated and competed. Ford and Chevrolet, two stalwarts of the American automotive landscape, recognized the imperative to expand beyond domestic borders to sustain growth and maintain market dominance. This chapter delves into the globalization strategies of Ford and Chevrolet, analyzing their expansion into international markets, the challenges they encountered, their successes, and how global competition influenced their domestic strategies.

The Dawn of Global Expansion

Why Globalization?

The push towards globalization for Ford and Chevrolet was driven by several factors:

- **Market Saturation in the U.S.:** As the domestic market matured, growth opportunities within the United States began to plateau.

- **Economic Opportunities Abroad:** Emerging markets offered untapped potential for sales and revenue.

- **Diversification of Risk:** Expanding into multiple regions helped mitigate the risks associated with economic downturns in any single market.

- **Economies of Scale:** Global operations allowed for cost reductions through larger production volumes and shared resources.

Strategic Objectives

Ford and Chevrolet aimed to achieve the following through globalization:

- **Increase Market Share:** Capture a significant portion of the global automotive market.

- **Brand Recognition:** Establish and strengthen their brands internationally.

- **Innovation and Learning:** Leverage global insights to drive innovation and improve product offerings.

- **Supply Chain Optimization:** Develop efficient global supply chains to enhance competitiveness.

Ford's Globalization Strategy

Early International Ventures

Ford Motor Company's Global Footprint

- **Europe:** Ford established a strong presence in Europe with the acquisition of brands like Ford of Britain and the development of models tailored to European tastes, such as the Ford Fiesta.

- **Asia:** Ford entered the Asian market through joint ventures and partnerships, notably with Mazda in Japan and later with Mahindra in India.

- **Latin America:** Ford expanded into Latin America, particularly in Brazil and Argentina, adapting models to suit local preferences and conditions.

Key Strategies Employed by Ford

Localization of Products

- **Adaptation to Local Tastes:** Ford tailored its vehicles to meet the specific needs and preferences of each market. For instance, the Ford Fiesta was designed with European consumers in mind, emphasizing fuel efficiency and compact dimensions suitable for narrow city streets.

- **Production in Local Markets:** Establishing manufacturing plants in key regions reduced transportation costs and allowed Ford to respond swiftly to local market demands. This strategy also fostered goodwill by contributing to local economies.

Strategic Alliances and Joint Ventures

- **Partnerships with Local Manufacturers:** Ford's collaboration with Mazda in Japan and with Mahindra in India exemplified its approach to leveraging local expertise and navigating regulatory landscapes.

- **Global Alliances:** Participation in global alliances, such as the Ford-Pinto partnership in South America, enabled Ford to share technology and resources across borders.

Global Branding and Marketing

- **Consistent Brand Messaging:** Ford maintained a consistent global brand image focused on reliability, performance, and innovation, while also allowing for localized marketing campaigns that resonated with regional audiences.

- **Sponsorships and Global Events:** Ford invested in

global sports and cultural events to enhance brand visibility and connect with diverse consumer bases.

Challenges Faced by Ford

Cultural Differences

- **Understanding Local Preferences:** Navigating the diverse cultural landscapes required Ford to deeply understand and respect local consumer behaviors and preferences, which sometimes led to missteps in product offerings.

- **Brand Perception:** Differing perceptions of American brands abroad necessitated careful management to build and maintain a positive brand image.

Economic and Political Barriers

- **Trade Regulations and Tariffs:** Varying trade policies, tariffs, and import restrictions posed significant barriers to entry in certain markets, increasing costs and complicating supply chains.

- **Political Instability:** In regions experiencing political turmoil, Ford had to navigate uncertainties that impacted operations and investments.

Competition from Local and Global Players

- **Local Manufacturers:** Established local competitors in markets like Europe and Asia presented formidable challenges, requiring Ford to continuously innovate and differentiate its products.

- **Global Automakers:** Competing against other global giants, such as Toyota and Volkswagen, demanded strategic agility and substantial investment in technology and marketing.

Ford's Successes in Global Markets

Europe: The Ford Fiesta and Focus

- **Ford Fiesta:** Launched in the late 1970s, the Fiesta became a cornerstone of Ford's European lineup, praised for its affordability, fuel efficiency, and adaptability.

- **Ford Focus:** Introduced in the late 1990s, the Focus garnered acclaim for its design, handling, and technology, cementing Ford's reputation in the compact car segment globally.

Asia: Strengthening Presence in Japan and India

- **Joint Ventures:** Ford's collaborations with Mazda and Mahindra facilitated entry into highly competitive and diverse markets, allowing for shared technological advancements and expanded product portfolios.

- **Local Adaptations:** Models like the Ford EcoSport tailored specifically for the Indian market showcased Ford's ability to innovate within different cultural contexts.

Latin America: Dominance in Brazil and Argentina

- **Ford Ka:** Designed for emerging markets, the Ford Ka achieved significant success in Brazil and Argentina, offering an affordable and efficient option that resonated with local consumers.

- **Robust Distribution Networks:** Establishing strong distribution and service networks ensured Ford's vehicles were accessible and supported across vast regions.

Chevrolet's Globalization Strategy

Chevrolet's International Expansion

Global Presence

- **Europe:** Chevrolet's presence in Europe was more limited compared to Ford, focusing on models like the Chevrolet Cobalt and later the Cruze to compete in the compact segment.

- **Asia:** Chevrolet expanded in markets like China and South Korea, often through joint ventures, to leverage local manufacturing and distribution networks.

- **Africa and Middle East:** Chevrolet targeted these regions with versatile models suited to varying terrains and climates, such as the Chevrolet Colorado and Silverado.

Key Strategies Employed by Chevrolet

Diversification of Product Lines

- **Wide Range of Models:** Chevrolet offered a diverse array of vehicles internationally, from compact cars and SUVs to trucks and performance models, catering to different market needs.

- **Specialized Vehicles for Specific Markets:** Models like the Chevrolet Sail in Asia and the Chevrolet Optra in Latin America were designed to meet unique regional demands.

Strategic Manufacturing and Supply Chain Management

- **Local Production Facilities:** Establishing manufacturing plants in key regions reduced production costs and allowed Chevrolet to customize vehicles for local markets effectively.

- **Efficient Supply Chains:** Implementing streamlined supply chain processes ensured timely delivery and reduced overheads, enhancing competitiveness.

Brand Localization and Marketing

- **Tailored Marketing Campaigns:** Chevrolet adapted its marketing strategies to align with local cultures and consumer behaviors, enhancing brand relevance and appeal.

- **Global and Local Branding Balance:** Maintaining a consistent global brand image while incorporating local elements ensured Chevrolet remained both recognizable and relatable.

Challenges Faced by Chevrolet

Limited Brand Recognition in Certain Markets

- **Competing Against Established Brands:** In regions dominated by other global and local automakers, Chevrolet had to work diligently to build brand recognition and trust.

- **Perception Issues:** Overcoming any negative perceptions associated with past products required strategic branding and high-quality offerings.

Economic and Political Instabilities

- **Market Volatility:** Economic fluctuations and political instability in emerging markets posed risks to Chevrolet's investments and operations.

- **Regulatory Compliance:** Navigating varying environmental and safety regulations across different regions required adaptability and compliance efforts.

Supply Chain and Production Complexities

- **Global Supply Chain Management:** Coordinating production and distribution across multiple regions introduced logistical complexities and potential inefficiencies.

- **Quality Control:** Ensuring consistent quality across diverse manufacturing facilities was a significant challenge, impacting brand reputation.

Chevrolet's Successes in Global Markets

Asia: Growth in China and South Korea

- **Chevrolet Cruze:** The Cruze achieved substantial success in China, offering a blend of affordability, efficiency, and modern features that appealed to a broad consumer base.

- **Local Partnerships:** Collaborations with local manufacturers in China facilitated smoother market entry and product localization, enhancing Chevrolet's competitive edge.

Latin America: Strong Presence in Brazil and Argentina

- **Chevrolet Onix:** Introduced in Brazil, the Onix

became one of the best-selling cars, praised for its fuel efficiency, affordability, and modern design.

- **Robust Product Lineup:** A diverse range of models, including the Chevrolet Silverado, catered to both personal and commercial users, ensuring broad market appeal.

Middle East and Africa: Versatile Offerings

- **Chevrolet Colorado and Silverado:** These models performed well in regions requiring durable and versatile trucks, leveraging Chevrolet's reputation for reliability and performance.

- **Adaptation to Local Needs:** Offering vehicles suited to diverse terrains and climates ensured Chevrolet remained relevant in challenging environments.

Influence of International Competition on Domestic Strategies

Enhanced Innovation and Technology Adoption

Global competition pushed Ford and Chevrolet to accelerate innovation and integrate advanced technologies into their domestic offerings:

- **Technology Transfer:** Innovations developed for international markets, such as fuel-efficient engines and advanced safety features, were adapted for domestic models, enhancing overall vehicle quality and performance.

- **Benchmarking Against Global Competitors:** Competing with international automakers like Toyota, Honda, and Volkswagen in global markets inspired Ford and Chevrolet to adopt best practices and improve their domestic product lines.

Cost Management and Efficiency Improvements

The need to compete globally drove both companies to optimize their cost structures and enhance production efficiencies:

- **Lean Manufacturing:** Implementing lean manufacturing principles reduced waste and improved operational efficiency, benefiting both international and domestic production.

- **Economies of Scale:** Global operations allowed for larger production volumes, lowering per-unit costs and enabling competitive pricing in the domestic market.

Product Diversification and Segmentation

Exposure to diverse international markets influenced Ford and Chevrolet to diversify and better segment their domestic product lines:

- **Expanding Vehicle Offerings:** Insights from global consumer preferences led to the introduction of new models and trims that catered to a wider

audience domestically.

- **Niche Markets:** Understanding the importance of niche markets abroad encouraged Ford and Chevrolet to develop specialized models and variants for specific domestic segments, such as performance enthusiasts and eco-conscious consumers.

Strengthened Brand Image and Reputation

Success and challenges in international markets impacted the brands' domestic reputation and consumer perception:

- **Global Brand Strength:** A strong presence internationally reinforced the brands' reliability and prestige at home, enhancing consumer trust and loyalty.

- **Reputation Management:** Failures or setbacks abroad necessitated robust brand management strategies to prevent negative spillover effects domestically.

Case Studies: Ford and Chevrolet's Global Ventures

Ford's Success in Europe: The Ford Fiesta and Focus

Ford Fiesta

- **Introduction and Market Penetration:** Launched in Europe in the late 1970s, the Fiesta quickly became a favorite due to its compact size, fuel efficiency, and affordability.

- **Cultural Adaptation:** Designed to suit European driving conditions and preferences, the Fiesta incorporated features like responsive handling and efficient engines.

- **Sales Performance:** The Fiesta consistently ranked among the best-selling cars in Europe, contributing significantly to Ford's international revenue.

Ford Focus

- **Launch and Evolution:** Introduced in the late 1990s,

the Focus was developed with extensive research into European market demands, emphasizing design, technology, and driving dynamics.

- **Critical Acclaim:** Garnered praise for its design, handling, and technological features, earning numerous awards and recognition in various international markets.

- **Market Impact:** The Focus strengthened Ford's position in Europe, competing effectively against established models from Volkswagen, Toyota, and other automakers.

Chevrolet's Triumph in Latin America: The Chevrolet Onix

Chevrolet Onix

- **Development and Launch:** Launched in Brazil in 2012, the Onix was designed specifically for the Latin American market, focusing on affordability, fuel efficiency, and modern features.

- **Market Penetration:** Rapidly became the best-selling car in Brazil, surpassing longstanding competitors due to its appealing design and value proposition.

- **Adaptation and Innovation:** Continuously updated with features like advanced infotainment systems and safety enhancements to maintain its competitive edge.

Impact on Chevrolet's Global Strategy

- **Replication in Other Markets:** Success with the Onix inspired Chevrolet to develop similar models for other emerging markets, leveraging the strategies that proved effective in Brazil.

- **Focus on Emerging Economies:** Reinforced Chevrolet's commitment to prioritizing growth in high-potential regions, influencing future investments and product development.

Ford's Foray into Asia: Joint Ventures and Local Adaptations

Ford-Mazda Alliance

- **Strategic Partnership:** Ford's alliance with Mazda facilitated entry into the Japanese market, allowing for shared technology and co-developed models.

- **Product Adaptations:** Collaborated on models like the Ford Probe, which blended American performance with Japanese engineering and design aesthetics.

- **Market Challenges:** Despite collaborative efforts, competition from established Japanese brands like Toyota and Honda remained intense, requiring continuous innovation and adaptation.

Ford in India: Partnership with Mahindra

- **Joint Ventures:** Ford partnered with Mahindra & Mahindra to produce and distribute vehicles tailored to the Indian market.

- **Localized Models:** Developed models like the Ford EcoSport, designed to meet the specific needs of Indian consumers, including compact dimensions and fuel-efficient engines.

- **Market Success:** Achieved notable sales growth by

addressing local preferences and leveraging Mahindra's established distribution network.

Chevrolet's Efforts in China and South Korea

Chevrolet in China

- **Market Entry:** Entered China through joint ventures, recognizing the importance of local partnerships in navigating regulatory and market complexities.

- **Localized Production:** Produced models like the Chevrolet Sail and Cruze in China, offering vehicles that matched local preferences for size, efficiency, and affordability.

- **Sales Growth:** Capitalized on the growing middle class and increasing automobile ownership rates in China, establishing a strong foothold in one of the world's largest automotive markets.

Chevrolet in South Korea

- **Strategic Alliances:** Collaborated with local manufacturers to produce and distribute vehicles tailored to South Korean consumers.

- **Model Adaptations:** Introduced models like the Chevrolet Spark, designed to appeal to urban drivers seeking compact and efficient vehicles.

- **Competitive Positioning:** Positioned Chevrolet as a provider of reliable and stylish vehicles, competing against both domestic and international brands.

Impact of International Competition on Domestic Strategies

Enhanced Product Innovation

Exposure to global competition necessitated greater emphasis on innovation within domestic markets:

- **Advanced Technologies:** Integration of technologies developed for international markets into domestic models, enhancing performance, safety, and efficiency.

- **Design Influences:** Incorporation of design elements and features inspired by global trends, ensuring domestic vehicles remained modern and appealing.

Cost Efficiency and Lean Operations

Global competition drove Ford and Chevrolet to optimize their domestic operations for greater cost efficiency:

- **Supply Chain Optimization:** Streamlined domestic supply chains by adopting best practices from international operations, reducing costs and improving quality.

- **Manufacturing Innovations:** Implemented lean manufacturing techniques and advanced production technologies to enhance productivity and reduce waste.

Diversification and Market Segmentation

Insights gained from international markets influenced Ford and Chevrolet to diversify and better segment their domestic offerings:

- **Expanded Vehicle Lineups:** Introduction of new models and trims to cater to diverse consumer preferences, mirroring successful strategies from global markets.

- **Targeted Marketing:** Developed targeted marketing campaigns for specific segments, such as eco-conscious consumers and performance enthusiasts, based on global market insights.

Strengthened Brand Positioning

Success and challenges in international markets influenced the brands' domestic positioning and reputation:

- **Global Brand Image:** A strong international presence reinforced the brands' reliability and prestige at home, enhancing consumer trust and loyalty.

- **Reputation Management:** Addressed any negative perceptions from international setbacks through robust domestic branding and quality improvement initiatives.

Challenges in Global Expansion

Cultural and Operational Barriers

Navigating diverse cultural landscapes and operational environments posed significant challenges:

- **Cultural Sensitivity:** Required deep understanding of local cultures to avoid missteps in marketing, product design, and customer engagement.

- **Operational Complexity:** Managing operations across multiple time zones, regulatory environments, and market conditions increased operational complexity and costs.

Economic and Political Instability

Operating in regions with economic volatility and political instability introduced risks:

- **Market Fluctuations:** Economic downturns and currency fluctuations affected profitability and sales performance in international markets.

- **Regulatory Changes:** Frequent changes in regulations required constant adaptation and compliance efforts, impacting operational efficiency.

Intense Competition from Local and Global Automakers

Competing against both local manufacturers and other global giants intensified the rivalry:

- **Local Competitors:** Established local brands with deep market knowledge and strong consumer loyalty presented formidable competition.

- **Global Peers:** Competing against other global automakers like Toyota, Honda, Volkswagen, and Hyundai required continuous innovation and strategic agility.

Supply Chain Disruptions

Global supply chains were vulnerable to disruptions caused by geopolitical tensions, natural disasters, and pandemics:

- **Logistical Challenges:** Coordinating production and distribution across multiple regions introduced logistical complexities and potential delays.

- **Resilience Building:** Required investment in supply chain resilience and diversification to mitigate the impact of disruptions.

Success Stories and Milestones

Ford's Triumphs Abroad

Ford Fiesta's European Dominance

- **Market Leadership:** The Fiesta consistently ranked among the best-selling cars in Europe, surpassing sales targets and outperforming competitors.

- **Adaptation Success:** Successfully adapted to European driving conditions and consumer preferences, featuring compact dimensions, efficient engines, and advanced safety features.

Ford Focus's Global Acclaim

- **Critical Success:** The Focus received numerous awards for design, safety, and performance, enhancing Ford's reputation in global markets.

- **Versatile Appeal:** Offered in a variety of body styles and configurations, the Focus appealed to a wide range of consumers, from urban drivers to performance enthusiasts.

Chevrolet's Global Achievements

Chevrolet Onix's Brazilian Success

- **Best-Selling Car:** The Onix became the best-selling car in Brazil, reflecting Chevrolet's ability to deliver a product that resonated with local consumers.

- **Continuous Improvement:** Regular updates and enhancements to the Onix ensured its continued appeal and competitiveness in the market.

Chevrolet Cruze's Chinese Growth

- **Strong Market Presence:** The Cruze gained significant market share in China, aided by strategic partnerships and localized product offerings.

- **Positive Reception:** Praised for its design, efficiency, and reliability, the Cruze solidified Chevrolet's position in one of the world's largest automotive markets.

Influence on Domestic Strategies

Product Development Synergies

Global insights influenced domestic product development, leading to:

- **Enhanced Features:** Introduction of advanced features and technologies initially developed for international markets into domestic models.

- **Design Refinements:** Adoption of design elements that proved successful globally, ensuring domestic vehicles remained competitive and modern.

Operational Best Practices

Implementing best practices from international operations improved domestic efficiency and quality:

- **Lean Manufacturing:** Adoption of lean principles from global operations reduced waste and improved production efficiency domestically.

- **Quality Control:** Enhanced quality control measures inspired by international standards elevated the overall quality of domestic vehicles.

Strategic Agility and Flexibility

Global competition fostered greater strategic agility and flexibility within Ford and Chevrolet:

- **Responsive Supply Chains:** Developed more responsive and flexible supply chains capable of adapting to changing market demands and conditions.

- **Innovative Marketing:** Leveraged global marketing strategies to enhance domestic campaigns, ensuring relevance and resonance with American consumers.

Risk Management and Diversification

Learning from global challenges informed better risk management and diversification strategies domestically:

- **Balanced Portfolios:** Diversified product portfolios to include a mix of fuel-efficient vehicles, performance models, and versatile trucks catering to varied consumer needs.

- **Geographic Diversification:** Continued emphasis on geographic diversification to spread risk and ensure stable revenue streams despite regional economic fluctuations.

Future Outlook: Sustaining Global Competitiveness

Embracing Electric and Autonomous Technologies

The ongoing shift towards electrification and autonomous

driving presents new opportunities and challenges:

- **Electric Vehicles (EVs):** Both Ford and Chevrolet are investing heavily in EV technology, developing electric versions of their iconic models like the Mustang Mach-E and the Chevrolet Bolt EV.

- **Autonomous Driving:** Research and development in autonomous driving technologies aim to position Ford and Chevrolet as leaders in the next generation of automotive innovation.

Expanding Presence in Emerging Markets

Continued focus on emerging markets remains crucial for sustained global growth:

- **Asia-Pacific Expansion:** Strengthening presence in high-growth regions like Southeast Asia and India through tailored product offerings and strategic partnerships.

- **Africa and Latin America:** Leveraging existing successes to deepen market penetration and explore new opportunities in these regions.

Sustainability and Environmental Responsibility

Commitment to sustainability is becoming increasingly important in global strategies:

- **Green Manufacturing:** Implementing sustainable manufacturing practices to reduce carbon footprints and promote environmental stewardship.

- **Eco-Friendly Products:** Developing a range of eco-friendly vehicles, including hybrids and fully electric models, to meet rising environmental standards and consumer demand.

Adapting to Regulatory Changes

Proactive adaptation to evolving regulatory landscapes ensures continued compliance and competitiveness:

- **Emission Standards:** Staying ahead of emission regulations by investing in cleaner technologies and efficient engine designs.

- **Safety Regulations:** Continuously enhancing safety features to meet and exceed international safety standards, reinforcing brand reputation for reliability and protection.

Leveraging Digital Transformation

Digital transformation is reshaping how automakers operate and engage with consumers:

- **Connected Vehicles:** Integrating advanced connectivity features into vehicles, offering enhanced infotainment, remote diagnostics, and over-the-air updates.

- **Digital Marketing and Sales:** Utilizing digital platforms for marketing, sales, and customer engagement to reach a broader and more tech-savvy audience.

Conclusion: Globalization as a Pillar of Growth and Innovation

The globalization efforts of Ford and Chevrolet have been instrumental in shaping their trajectories as leading automotive manufacturers. By expanding into international markets, both companies have unlocked new avenues for growth, driven innovation, and reinforced their brand identities on a global scale. While the journey has been fraught with challenges, the successes achieved in diverse regions underscore the resilience and strategic agility of these automotive giants.

Global competition has not only influenced their international strategies but has also driven significant advancements and improvements in their domestic operations. The lessons learned from navigating different cultural, economic, and regulatory landscapes have fostered a culture of continuous improvement and innovation within both companies.

As the automotive industry continues to evolve with the advent of electric vehicles, autonomous technologies, and shifting consumer preferences, the foundation laid through globalization will be pivotal in sustaining the competitiveness and relevance of Ford and Chevrolet. Embracing these changes with the same spirit of innovation and adaptability that defined their early globalization efforts will ensure that these iconic brands remain at the forefront of the global automotive arena.

The rivalry between Ford and Chevrolet, now expanded onto the global stage, exemplifies how competition can drive progress, foster innovation, and shape the future of an entire industry. As they navigate the complexities of an interconnected world, Ford and Chevrolet continue to exemplify the enduring legacy and dynamic evolution of American automotive excellence.

Chapter 14: The Digital Age—Marketing and Community Building

As the turn of the millennium ushered in unprecedented technological advancements, the automotive industry underwent a profound transformation. The advent of the internet and the rise of social media revolutionized how manufacturers like Ford and Chevrolet interacted with their customers, shaped their brand images, and fostered loyalty. Concurrently, the proliferation of online communities, car clubs, forums, and events played a pivotal role in strengthening the bond between enthusiasts and their preferred brands. This chapter delves into the multifaceted impact of the digital age on Ford and Chevrolet, exploring how these automotive giants harnessed digital tools to enhance marketing strategies and build vibrant, engaged communities.

The Digital Revolution: Transforming Automotive Marketing

The Internet Era: A New Frontier for Marketing

The emergence of the internet in the late 20th century opened new avenues for automotive marketing. For Ford and Chevrolet, the internet became a crucial platform to reach broader audiences, provide detailed product information, and engage with customers in real-time.

Website Development and Online Presence

- **Official Websites:** Both Ford and Chevrolet invested heavily in developing comprehensive websites that served as hubs for product information, dealer locators, financing options, and customer support. These websites became the first point of contact for potential buyers, offering interactive tools like virtual tours, build-and-price features, and multimedia content.

- **E-Commerce Integration:** Introducing online purchasing options and facilitating seamless transactions allowed customers to explore and purchase vehicles from the comfort of their homes, enhancing convenience and accessibility.

Search Engine Optimization (SEO) and Online Advertising

- **SEO Strategies:** Optimizing website content for search engines ensured that Ford and Chevrolet remained visible to consumers searching for vehicles, increasing organic traffic and lead generation.

- **Pay-Per-Click (PPC) Campaigns:** Targeted online advertising through PPC campaigns enabled both brands to reach specific demographics, track campaign performance, and adjust strategies in real-time for maximum effectiveness.

Social Media: Shaping Brand Image and Customer Engagement

The rise of social media platforms like Facebook, Twitter, Instagram, and YouTube transformed the way Ford and Chevrolet interacted with their audiences. Social media provided a dynamic environment for brand storytelling, real-time communication, and community engagement.

Content Marketing and Storytelling

- **Engaging Content:** Both brands utilized social media to share engaging content, including behind-the-scenes looks at manufacturing, interviews with designers and engineers, and stories highlighting the heritage and innovation of their vehicles.

- **Visual Storytelling:** Platforms like Instagram and YouTube allowed Ford and Chevrolet to showcase high-quality visuals and videos, emphasizing the design, performance, and lifestyle aspects of their models.

Real-Time Communication and Customer Service

- **Direct Interaction:** Social media enabled direct interaction between the brands and their customers. Responding to inquiries, addressing concerns, and engaging in conversations helped build trust and loyalty.

- **Crisis Management:** In instances of product recalls or negative publicity, timely and transparent communication on social media platforms allowed both companies to manage crises effectively and maintain customer trust.

Influencer Partnerships and Sponsored Content

- **Collaborations with Influencers:** Partnering with automotive influencers, bloggers, and celebrities expanded the reach of Ford and Chevrolet's marketing efforts. Influencers provided authentic reviews, test drives, and endorsements that resonated with their followers.

- **Sponsored Campaigns:** Creating sponsored content and advertisements on social media platforms helped target specific audiences, promoting new models, special editions, and seasonal offers.

Analytics and Data-Driven Marketing

The digital age empowered Ford and Chevrolet with advanced analytics tools to monitor and analyze customer behavior, campaign performance, and market trends.

Customer Insights and Personalization

- **Data Collection:** Gathering data from website interactions, social media engagement, and online purchases provided valuable insights into customer preferences and behaviors.

- **Personalized Marketing:** Utilizing data-driven insights allowed both brands to tailor marketing messages, offers, and recommendations to individual customers, enhancing the relevance and effectiveness of their campaigns.

Performance Tracking and Optimization

- **Campaign Analytics:** Monitoring key performance indicators (KPIs) such as click-through rates, conversion rates, and engagement metrics enabled Ford and Chevrolet to evaluate the success of their marketing efforts.

- **Continuous Improvement:** Analyzing campaign data facilitated ongoing optimization, allowing for adjustments in real-time to improve ROI and achieve marketing objectives.

Fan Communities: The Heartbeat of Brand Loyalty

Beyond digital marketing, the rise of online communities, car clubs, forums, and events played a critical role in fostering brand loyalty and creating a sense of belonging among enthusiasts. Ford and Chevrolet leveraged these platforms to build strong, engaged communities around their iconic models—the Mustang and Camaro.

Online Forums and Discussion Boards

Online forums and discussion boards became virtual meeting places for Ford and Chevrolet enthusiasts to share experiences, seek advice, and celebrate their vehicles.

Official and Unofficial Forums

- **Official Forums:** Both Ford and Chevrolet established official forums where owners could interact with company representatives, access exclusive content, and participate in brand-sponsored events.

- **Unofficial Communities:** Independent forums like Mustang6G.com and Camaro5.com provided spaces for passionate fans to engage in in-depth discussions, modifications, and sharing of customization ideas.

Knowledge Sharing and Support

- **Technical Support:** Forums served as valuable resources for troubleshooting, maintenance tips, and technical support, enhancing the ownership experience.

- **Modification and Customization:** Enthusiasts shared modification projects, customization tips, and performance upgrades, fostering a culture of creativity and innovation within the community.

Car Clubs and Enthusiast Groups

Car clubs and enthusiast groups brought together Ford and Chevrolet owners for in-person gatherings, drives, and events, strengthening the bond between the brands and their customers.

Official and Regional Clubs

- **Ford Mustang Club of America (FMCA):** One of the largest and most active Mustang clubs, the FMCA organizes regional and national events, rallies, and meet-ups, providing members with exclusive access to events and merchandise.

- **Chevrolet Camaro Club of America (CCCA):** Similar to the FMCA, the CCCA facilitates gatherings, track days, and social events for Camaro enthusiasts, fostering a strong sense of community.

Exclusive Membership Benefits

- **Access to Events:** Members of official car clubs enjoy priority access to exclusive events, such as product launches, track days, and social gatherings.

- **Merchandise and Discounts:** Clubs often offer branded merchandise, discounts on parts and accessories, and special offers on vehicles, enhancing member loyalty and satisfaction.

Events and Shows: Celebrating Heritage and Innovation

Automotive events and shows provided platforms for Ford and Chevrolet to showcase their latest models, celebrate their heritage, and engage directly with fans.

Auto Shows and Expos

- **Major International Auto Shows:** Participation in events like the North American International Auto Show (NAIAS) allowed Ford and Chevrolet to unveil new models, demonstrate technological advancements, and interact with a global audience.

- **Local and Regional Shows:** Engaging with local car shows and expos helped both brands connect with communities on a more personal level, showcasing vehicles tailored to regional preferences.

Track Days and Performance Events

- **Performance Demonstrations:** Organizing track days and performance events enabled enthusiasts to experience the capabilities of their vehicles firsthand, fostering excitement and appreciation for the brand's engineering prowess.

- **Motorsports Integration:** Leveraging successes in motorsports, such as Mustang's racing heritage and Camaro's drag racing legacy, Ford and Chevrolet integrated motorsports elements into their events, appealing to performance-oriented fans.

Digital and Virtual Communities

The digital age also gave rise to virtual communities, where fans could connect, share content, and engage with the brands in innovative ways.

Social Media Groups and Pages

- **Facebook Groups:** Dedicated Facebook groups for Mustang and Camaro enthusiasts provided platforms for sharing photos, organizing meet-ups, and discussing vehicle-related topics.

- **Instagram and Twitter Communities:** Hashtags, dedicated pages, and interactive posts on Instagram and Twitter fostered real-time engagement and showcased the vibrant lifestyles associated with the brands.

Virtual Events and Webinars

- **Live Streams and Webinars:** Hosting live streams of events, Q&A sessions with designers and engineers, and webinars on vehicle maintenance and customization kept the community engaged and informed.

- **Online Contests and Challenges:** Engaging fans through online contests, photo challenges, and virtual car shows encouraged participation and amplified the brands' online presence.

Case Studies: Successful Digital and Community Strategies

Ford's Digital Engagement: Mustang's Online Presence

Social Media Campaigns

- **#MustangHeritage:** A campaign celebrating the Mustang's history through archival photos, stories, and fan-submitted content, fostering a sense of pride and connection among owners.

- **Interactive Features:** Utilizing Instagram Stories, Facebook Live, and Twitter polls to engage fans in real-time, gather feedback, and involve them in the brand's narrative.

Official Mustang Website Enhancements

- **Interactive Tools:** Incorporating build-and-price tools, virtual reality tours, and configurators that allowed customers to personalize their Mustangs online.

- **Content Hubs:** Creating dedicated sections for

Mustang news, events, and community stories to keep enthusiasts informed and connected.

Chevrolet's Community Building: Camaro's Enthusiast Networks

Camaro Clubs and Ambassador Programs

- **Official Camaro Ambassador Program:** Selecting passionate owners to represent the brand at events, online, and in media, fostering authentic brand advocacy.

- **Regional Camaro Clubs:** Supporting local clubs with resources, sponsorships, and exclusive content to strengthen community bonds.

Digital Content and Influencer Partnerships

- **YouTube Series:** Launching series like "Camaro Chronicles" featuring restoration projects, performance tests, and interviews with owners.

- **Influencer Collaborations:** Partnering with automotive influencers to create authentic content, reviews, and promotional material that resonated with younger audiences.

Interactive Platforms: Engaging the Next Generation

Mobile Applications

- **FordPass and Chevrolet MyLink:** Developing mobile apps that offer vehicle management tools, access to exclusive content, and enhanced connectivity features, integrating the digital experience with vehicle ownership.

- **Gamification Elements:** Incorporating gameral features such as loyalty points, achievement badges for community participation, and interactive vehicle customization options to keep users engaged and motivated.

Augmented Reality (AR) Experiences

- **Virtual Showrooms:** Utilizing AR to allow customers to explore and interact with vehicle models in a virtual environment, enhancing the online shopping experience.

- **Interactive Maintenance Guides:** Providing AR-based tutorials for vehicle maintenance and customization, making it easier for enthusiasts to work on their cars.

Impact of Digital and Community Strategies on Brand Loyalty

Enhanced Customer Experience

The integration of digital tools and community engagement strategies significantly enhanced the overall customer experience, making it more personalized, interactive, and enjoyable.

- **Seamless Interaction:** Customers could engage with the brands across multiple touchpoints, from online platforms to in-person events, creating a cohesive and consistent brand experience.

- **Personalized Engagement:** Tailored content, personalized marketing messages, and interactive tools made customers feel valued and understood, fostering deeper emotional connections with the brands.

Strengthened Brand Image

Effective digital marketing and community-building efforts reinforced the positive brand image of Ford and Chevrolet, positioning them as modern, innovative, and customer-centric.

- **Modern Appeal:** Embracing digital trends and technologies kept the brands relevant and appealing to younger, tech-savvy consumers.

- **Heritage Preservation:** Celebrating the rich histories of the Mustang and Camaro through digital content and community events honored the brands' legacies while showcasing their evolution.

Increased Customer Loyalty and Advocacy

Engaged communities and interactive digital strategies turned satisfied customers into loyal advocates who actively promoted the brands through word-of-mouth and social media.

- **Brand Ambassadors:** Enthusiastic community members and official ambassadors became unofficial brand representatives, amplifying the brands' reach and influence.

- **Recurring Engagement:** Continuous interaction through events, online content, and community activities ensured that customers remained engaged and loyal over time.

Feedback and Continuous Improvement

Digital platforms provided valuable channels for collecting customer feedback, allowing Ford and Chevrolet to continuously refine their products and strategies based on real-time insights.

- **Customer Insights:** Analyzing engagement data and feedback from online communities helped identify trends, preferences, and areas for improvement.

- **Agile Adaptation:** The ability to quickly respond to customer needs and market changes enabled both brands to stay ahead of the competition and maintain customer satisfaction.

Challenges in Digital Marketing and Community Building

Managing Online Reputation

Maintaining a positive online reputation required constant vigilance and proactive management to address negative comments, misinformation, and potential PR crises.

- **Active Monitoring:** Implementing robust social media monitoring tools to track brand mentions, sentiment, and emerging issues.

- **Responsive Communication:** Establishing clear protocols for responding to customer concerns, complaints, and negative feedback in a timely and effective manner.

Balancing Personalization and Privacy

While personalization enhances customer engagement, it also raises concerns about data privacy and security.

- **Data Protection:** Ensuring compliance with data protection regulations like GDPR and CCPA by implementing stringent data privacy measures.

- **Transparent Practices:** Communicating clearly with customers about data collection practices, usage, and protection to build trust and transparency.

Content Overload and Engagement Fatigue

With the abundance of digital content, maintaining high levels of engagement without overwhelming the audience posed a significant challenge.

- **Quality over Quantity:** Focusing on creating high-quality, valuable content that resonates with the audience rather than flooding channels with excessive information.

- **Strategic Content Planning:** Developing a well-thought-out content strategy that includes a balanced mix of informational, entertaining, and promotional content to keep the audience engaged and interested.

Resource Allocation and Management

Effective digital marketing and community-building require substantial resources, including time, personnel, and budget.

- **Dedicated Teams:** Establishing specialized teams to manage digital marketing, social media, and community engagement efforts, ensuring focused and efficient operations.

- **Continuous Training:** Investing in ongoing training and development for team members to stay updated with the latest digital trends, tools, and best practices.

Future Trends: The Evolving Digital Landscape

Artificial Intelligence (AI) and Machine Learning

AI and machine learning are set to revolutionize digital marketing and community engagement for Ford and Chevrolet.

- **Predictive Analytics:** Utilizing AI to predict customer behaviors, preferences, and trends, enabling more targeted and effective marketing campaigns.

- **Chatbots and Virtual Assistants:** Implementing AI-driven chatbots to provide instant customer support, answer queries, and facilitate vehicle customization processes.

Virtual Reality (VR) and Immersive Experiences

VR technology offers immersive experiences that can transform how customers interact with Ford and Chevrolet brands.

- **Virtual Test Drives:** Allowing customers to experience virtual test drives of vehicles, enhancing the online shopping experience and reducing the need for physical visits.

- **Immersive Showrooms:** Creating virtual showrooms where customers can explore vehicle features, design options, and performance capabilities in a fully interactive environment.

Blockchain and Enhanced Security

Blockchain technology can enhance data security and transparency in digital marketing and community interactions.

- **Secure Transactions:** Utilizing blockchain for secure and transparent vehicle transactions, financing, and ownership records.

- **Verified Reviews and Content:** Implementing blockchain to verify the authenticity of customer reviews and community-generated content, ensuring trust and reliability.

Sustainability and Eco-Friendly Initiatives

As environmental consciousness grows, digital marketing and community strategies will increasingly emphasize sustainability and eco-friendly initiatives.

- **Green Marketing:** Highlighting sustainable practices, eco-friendly vehicle options, and corporate social responsibility (CSR) efforts through digital channels.

- **Community Engagement in Sustainability:** Encouraging community participation in sustainability initiatives, such as car-sharing programs, recycling drives, and environmental advocacy.

Conclusion: Embracing the Digital Future with Community at the Core

The digital age has redefined the landscape of automotive marketing and community building, offering unparalleled opportunities for brands like Ford and Chevrolet to connect with their audiences, enhance their brand images, and foster unwavering customer loyalty. By embracing the power of the internet and social media, and by nurturing vibrant communities through car clubs, forums, and events, Ford and Chevrolet have successfully navigated the complexities of the digital era.

The integration of advanced technologies, personalized marketing strategies, and community-centric initiatives has not only strengthened the bond between the brands and their customers but has also set new standards for engagement and loyalty in the automotive industry. As technology continues to evolve, the ability to adapt and innovate will remain paramount, ensuring that Ford and Chevrolet continue to lead the way in building lasting, meaningful relationships with their global communities.

The revival of the classics in the digital age exemplifies how honoring heritage while embracing modernity can create a powerful synergy that resonates with both long-time enthusiasts and new generations of drivers. As we look to the future, the principles of digital engagement and community building will continue to be essential in sustaining the enduring legacy and dynamic evolution of Ford and Chevrolet.

Chapter 15: Innovations in Sustainability and Technology

In the face of escalating environmental concerns and rapid technological advancements, the automotive industry stands at a pivotal crossroads. Ford and Chevrolet, two of America's most iconic automakers, are spearheading initiatives to embrace sustainability and integrate cutting-edge technologies into their vehicle lineups. This chapter explores the recent developments in electric vehicles (EVs), autonomous driving, and sustainable manufacturing practices undertaken by Ford and Chevrolet. Additionally, it examines how these innovations are positioning both brands for a competitive edge in the future automotive landscape.

Electric Vehicles: Pioneering the Green Revolution

Ford's Electric Vehicle Portfolio

Ford Mustang Mach-E

- **Overview**: Launched in 2020, the Mustang Mach-E marks Ford's significant entry into the electric SUV market, blending the heritage of the Mustang with modern EV technology.

- **Specifications**:

    - **Battery Options**: Available in Standard Range (68 kWh) and Extended Range (88 kWh) configurations.

    - **Performance**: Ranges from 266 horsepower in the Select trim to 480 horsepower in the GT trim.

    - **Range**: Up to 305 miles on a single charge (Extended Range, RWD).

- **Market Impact**: The Mach-E received critical acclaim for its performance, design, and technology, positioning Ford as a serious contender in the competitive EV market.

Ford F-150 Lightning

- **Overview**: Introduced in 2021, the F-150 Lightning is Ford's first all-electric version of its best-selling F-Series truck.

- **Specifications**:

    - **Powertrain**: Dual motors delivering up to 563 horsepower and 775 lb-ft of torque.

    - **Range**: Approximately 230 miles on a single charge.

    - **Features**: Pro Power Onboard system providing up to 9.6 kW of power for tools and appliances, integrated Ford Co-Pilot360 for advanced driver assistance.

- **Market Impact**: The F-150 Lightning has been lauded for maintaining the utility and performance of the traditional F-150 while introducing innovative electric features, appealing to both traditional truck buyers and new EV enthusiasts.

Ford E-Transit

- **Overview**: An all-electric version of Ford's popular Transit van, launched to meet the growing demand for sustainable commercial vehicles.

- **Specifications**:

    o **Battery Capacity**: 67 kWh, providing a range of up to 126 miles.

    o **Payload**: Capable of handling the same payload as its internal combustion counterpart.

    o **Features**: Advanced connectivity options, flexible interior configurations for various business needs.

- **Market Impact**: The E-Transit addresses the commercial sector's need for eco-friendly transportation solutions, enhancing Ford's position in the electric commercial vehicle market.

Chevrolet's Electric Vehicle Portfolio

Chevrolet Bolt EV and Bolt EUV

- **Overview**: The Bolt EV and its crossover counterpart, the Bolt EUV, represent Chevrolet's efforts to provide affordable, practical electric vehicles.

- **Specifications**:

    o **Battery Capacity**: 66 kWh, offering an EPA-estimated range of up to 259 miles (Bolt EV) and 247 miles (Bolt EUV).

    o **Performance**: Front-wheel drive with a single motor producing 200 horsepower.

    o **Features**: Super Cruise hands-free driving technology (Bolt EUV), spacious interior, and advanced infotainment systems.

- **Market Impact**: The Bolt series has been praised for its affordability, range, and practicality, making EVs more accessible to a broader audience and reinforcing Chevrolet's commitment to electric mobility.

Chevrolet Silverado EV

- **Overview**: Announced in 2022, the Silverado EV is Chevrolet's all-electric version of its flagship Silverado truck.

- **Specifications**:
    - **Powertrain**: Dual electric motors with up to 664 horsepower and 780 lb-ft of torque.
    - **Range**: Over 400 miles on a single charge, depending on configuration.
    - **Features**: Ultra-quiet operation, advanced towing capabilities, innovative battery technology for quick charging.

- **Market Impact**: The Silverado EV aims to compete directly with Ford's F-150 Lightning, offering superior range and performance, thereby strengthening Chevrolet's position in the electric truck market.

Chevrolet Blazer EV

- **Overview**: A recently introduced all-electric SUV that combines sporty design with eco-friendly performance.

- **Specifications**:
    - **Battery Capacity**: 100 kWh, providing an

estimated range of up to 400 miles.

- **Performance**: Multiple powertrain options, including dual-motor all-wheel drive configurations.

- **Features**: Advanced driver-assistance systems, luxurious interior materials, and cutting-edge infotainment technology.

- **Market Impact**: The Blazer EV caters to consumers seeking a blend of style, performance, and sustainability, enhancing Chevrolet's EV lineup with a versatile and attractive option.

Autonomous Driving: Steering Towards the Future

Ford's Autonomous Driving Initiatives

Ford BlueCruise

- **Overview**: Ford's hands-free driving technology designed for compatible highways, enabling drivers to relax their hands on the wheel while maintaining control.

- **Features**:

  - **Hands-Free Capability**: Allows hands-free driving on pre-mapped, divided highways.

  - **Driver Monitoring System**: Ensures driver attention and readiness to take over when necessary.

  - **Integration with Ford Sync**: Seamlessly integrates with Ford's infotainment system for enhanced user experience.

- **Market Impact**: BlueCruise positions Ford as a leader in semi-autonomous driving technology, enhancing safety and convenience for drivers.

Ford Autonomous Vehicles (AV) Research

- **Overview**: Ford has invested significantly in autonomous vehicle research and development, partnering with companies like Argo AI to advance self-driving technologies.

- **Projects**:
  - **Argo AI Partnership**: Collaborative efforts to develop Level 4 autonomous driving systems for urban environments.
  - **Test Fleets**: Deployment of autonomous test fleets in select cities to refine and validate AV technologies.

- **Market Impact**: Ford's commitment to AV research underscores its dedication to leading the future of transportation, aiming to offer fully autonomous vehicles in the coming years.

Chevrolet's Autonomous Driving Initiatives

Chevrolet Super Cruise

- **Overview**: An advanced hands-free driving assistance system available on select Chevrolet models, including the Bolt EUV and Silverado EV.

- **Features**:
  - **Hands-Free Operation**: Enables hands-free driving on compatible highways with real-time

lane-keeping and adaptive cruise control.

- **Driver Attention Monitoring**: Utilizes eye-tracking technology to ensure driver engagement.

- **Over-the-Air Updates**: Continuously improves system performance and features through software updates.

- **Market Impact**: Super Cruise enhances Chevrolet's competitive edge in driver-assistance technologies, offering a premium feature that appeals to safety-conscious consumers.

Chevrolet Autonomous Vehicle Research

- **Overview**: Chevrolet is actively involved in the development of autonomous driving technologies through partnerships and in-house research.

- **Projects**:

  - **Collaboration with Autonomous Tech Firms**: Partnering with companies like Cruise and Waymo to explore Level 4 and Level 5 autonomous driving capabilities.

  - **Pilot Programs**: Implementing pilot programs in urban areas to test and refine autonomous systems.

- **Market Impact**: Chevrolet's proactive approach to autonomous technology ensures its readiness to compete in the future of self-driving vehicles, aiming to integrate fully autonomous features into its lineup.

Sustainable Manufacturing: Building a Greener Future

Ford's Sustainable Manufacturing Practices

Commitment to Carbon Neutrality

- **Goals**: Ford aims to achieve carbon neutrality globally by 2050, with interim targets to reduce emissions and increase sustainability across its operations.

- **Strategies**:

    - **Renewable Energy**: Transitioning manufacturing facilities to 100% renewable energy sources.

    - **Energy Efficiency**: Implementing energy-efficient technologies and processes to minimize energy consumption.

    - **Carbon Offsetting**: Investing in carbon offset projects to compensate for unavoidable emissions.

Circular Economy Initiatives

- **Overview**: Ford embraces the principles of the circular economy to reduce waste and promote the reuse and recycling of materials.

- **Programs**:

    - **Vehicle Recycling Programs**: Ensuring end-of-life vehicles are dismantled responsibly, with materials recycled or repurposed.

    - **Sustainable Materials**: Increasing the use of

recycled and sustainable materials in vehicle production, such as recycled plastics and sustainable textiles.

- **Market Impact**: These initiatives not only reduce Ford's environmental footprint but also appeal to eco-conscious consumers, enhancing brand reputation.

Water and Waste Management

- **Overview**: Ford has implemented comprehensive water and waste management systems to minimize environmental impact.

- **Strategies**:

    o **Water Recycling**: Advanced water recycling systems in manufacturing plants reduce freshwater usage.

    o **Waste Reduction**: Streamlining production processes to minimize waste generation and maximize recycling rates.

- **Market Impact**: Effective water and waste management contribute to Ford's sustainability goals, ensuring responsible resource usage and compliance with environmental regulations.

Chevrolet's Sustainable Manufacturing Practices

General Motors Sustainability Goals

- **Overview**: As part of General Motors (GM), Chevrolet benefits from the company's overarching sustainability initiatives aimed at reducing environmental impact.

- **Goals**:
    - **Carbon Neutrality by 2040**: GM aims to be carbon neutral across its global operations by 2040.
    - **Zero Waste to Landfill**: Achieving zero waste to landfill in all manufacturing facilities.
    - **Sustainable Sourcing**: Ensuring that all materials are sustainably sourced and responsibly managed.

Energy Efficiency and Renewable Energy

- **Overview**: Chevrolet's manufacturing plants incorporate energy-efficient technologies and prioritize the use of renewable energy sources.
- **Strategies**:
    - **Solar Power Installations**: Installing solar panels at key facilities to generate clean energy and reduce reliance on fossil fuels.
        - **LED Lighting and Efficient HVAC Systems**: Upgrading to LED lighting and implementing energy-efficient HVAC systems to lower energy consumption.

Sustainable Materials and Recycling

- **Overview**: Chevrolet emphasizes the use of sustainable materials and robust recycling programs to minimize environmental impact.

- **Initiatives**:
    - **Recycled and Bio-Based Materials**: Incorporating recycled plastics, bio-based textiles, and other sustainable materials into vehicle interiors and components.
        - **Comprehensive Recycling Programs**: Ensuring that manufacturing waste is thoroughly recycled or repurposed, reducing landfill contributions.

Water Conservation and Pollution Prevention

- **Overview**: Chevrolet implements advanced water conservation and pollution prevention measures in its manufacturing processes.

- **Strategies**:
    - **Closed-Loop Water Systems**: Utilizing closed-loop systems to recycle and reuse water within manufacturing plants.
        - **Pollution Control Technologies**: Installing advanced filtration and treatment systems to prevent pollutants from entering natural water sources.

Competitive Edge: Positioning for the Future

Ford's Strategic Positioning

Leadership in Electric and Autonomous Vehicles

- **Comprehensive EV Lineup**: With models like the Mustang Mach-E, F-150 Lightning, and E-Transit, Ford offers a diverse range of electric vehicles catering to different market segments.

- **Autonomous Driving Pioneering**: Investments in Ford BlueCruise and partnerships with Argo AI position Ford as a leader in autonomous driving technologies.

Sustainability as a Core Value

- **Carbon Neutral Goals**: Ford's commitment to carbon neutrality by 2050 and ongoing sustainable manufacturing practices enhance its appeal to eco-conscious consumers.

- **Circular Economy Initiatives**: Embracing circular economy principles ensures long-term sustainability and resource efficiency, strengthening Ford's market position.

Technological Integration and Innovation

- **Advanced Connectivity**: Features like FordPass and SYNC integrate seamlessly with consumers' digital lifestyles, offering enhanced convenience and connectivity.

- **Pro Power Onboard**: Innovative features like Pro Power Onboard in the F-150 Lightning provide unique value propositions, differentiating Ford's EV offerings from competitors.

Chevrolet's Strategic Positioning

Expansive Electric Vehicle Portfolio

- **Diverse EV Offerings**: With the Bolt EV, Silverado EV, and upcoming models like the Blazer EV, Chevrolet caters to a wide range of consumer needs, from compact cars to full-size trucks.

- **Performance and Range**: Chevrolet's focus on performance and long-range capabilities in its EV models ensures competitiveness in the high-performance EV market.

Advanced Autonomous Driving Technologies

- **Super Cruise Integration**: Super Cruise technology on the Bolt EUV and Silverado EV offers advanced driver-assistance features, enhancing safety and convenience.

- **Collaborative Research**: Partnerships with autonomous tech firms enable Chevrolet to stay at the forefront of self-driving technology development.

Commitment to Sustainable Manufacturing

- **General Motors' Sustainability Goals**: Aligning with GM's ambitious sustainability targets, Chevrolet benefits from comprehensive initiatives aimed at reducing environmental impact.

- **Eco-Friendly Materials and Processes**: Emphasizing the use of sustainable materials and efficient manufacturing processes reinforces Chevrolet's dedication to environmental responsibility.

Market Differentiation and Brand Strength

Ford's Brand Strength

- **Heritage and Innovation**: Combining the rich heritage of the Mustang and F-Series with modern electric and autonomous technologies creates a powerful brand narrative.

- **Community and Loyalty**: Strong fan communities and active engagement through digital platforms foster deep brand loyalty and advocacy.

Chevrolet's Brand Strength

- **Versatility and Performance**: Offering a versatile range of vehicles that balance performance, comfort, and sustainability appeals to a broad consumer base.

- **Global Reach and Adaptability**: Chevrolet's successful global expansion and ability to adapt products for diverse markets enhance its brand strength and market presence.

Innovation and Continuous Improvement

- **Research and Development Investment**: Both Ford and Chevrolet's significant investments in R&D drive continuous innovation, ensuring their offerings remain competitive and cutting-edge.

- **Feedback-Driven Enhancements**: Leveraging customer feedback and data analytics allows both brands to refine their products and services, enhancing customer satisfaction and loyalty.

Conclusion: Steering Towards a Sustainable and Technological Future

The innovations in sustainability and technology undertaken by Ford and Chevrolet are pivotal in shaping the future of the automotive industry. By embracing electric mobility, advancing autonomous driving capabilities, and committing to sustainable manufacturing practices, both brands are not only addressing current environmental and technological challenges but also positioning themselves as leaders in the next generation of automotive excellence.

Ford's comprehensive approach, combining a diverse electric vehicle lineup, pioneering autonomous technologies, and steadfast sustainability goals, ensures its resilience and adaptability in a rapidly evolving market. Meanwhile, Chevrolet's expansive electric portfolio, advanced driver-assistance systems, and alignment with General Motors' sustainability initiatives solidify its competitive edge and commitment to a greener, more technologically integrated future.

As the automotive landscape continues to transform with emerging technologies and shifting consumer priorities, the strategic innovations of Ford and Chevrolet will play a crucial role in determining their continued dominance and relevance. Their ability to innovate, adapt, and lead in sustainability and technology not only secures their market positions but also contributes to the broader movement towards a sustainable and technologically advanced transportation ecosystem.

The rivalry between Ford and Chevrolet, now extended into the realms of sustainability and cutting-edge technology, exemplifies how competition can drive progress and foster a commitment to excellence. As they navigate the challenges and opportunities of the digital age, Ford and Chevrolet remain steadfast in their pursuit of automotive innovation, ensuring their enduring legacy and continued influence in the global automotive industry.

Chapter 16: The Impact on Pop Culture

Automobiles have long transcended their utilitarian purpose, evolving into symbols of freedom, power, and identity. Among these, Ford and Chevrolet have cemented their places not only on roads but also in the collective imagination through their significant presence in movies, music, and art. This chapter explores how these iconic brands have influenced and been influenced by pop culture, examining their appearances in various media, the creation of enduring symbols, and the symbiotic relationship between automotive excellence and cultural expression.

Automobiles as Cultural Icons

Symbolism and Identity

Automobiles often serve as extensions of personal identity, embodying the aspirations, lifestyles, and values of their owners. Ford and Chevrolet, with their diverse lineups ranging from rugged trucks to sleek sports cars, have provided vehicles that cater to a wide spectrum of cultural archetypes.

- **Freedom and Adventure:** Vehicles like the Ford Mustang and Chevrolet Camaro represent freedom and the spirit of adventure, appealing to those who seek excitement and individuality.

- **Power and Performance:** The dominance of Ford's F-Series and Chevrolet's Silverado in the truck segment symbolizes strength, reliability, and the hardworking ethos of American culture.

- **Innovation and Modernity:** Electric models like the Ford Mustang Mach-E and Chevrolet Bolt EV reflect a commitment to innovation and sustainability, resonating with environmentally conscious consumers.

Automobiles in Storytelling

Cars are integral to storytelling in various media, often serving as key plot devices or symbols that enhance narratives.

- **Character Development:** Vehicles can define characters, reflecting their personalities, statuses, and transformations. A protagonist's choice of car can signify their journey, challenges, and growth.

- **Plot Devices:** Cars can drive the plot forward, whether through high-speed chases, heroic rescues, or pivotal moments of conflict and resolution.

- **Emotional Connections:** Iconic cars evoke nostalgia and emotional responses, creating lasting impressions that resonate with audiences across generations.

Presence in Movies

Ford Mustang in "Bullitt" (1968)

One of the most iconic car appearances in film history, the Ford Mustang GT 390 driven by Steve McQueen in "Bullitt" set a new standard for automotive presence in cinema.

- **High-Speed Chase Scenes:** The Mustang's performance was showcased in exhilarating car chases through the streets of San Francisco, highlighting its speed, handling, and rugged design.

- **Cultural Impact:** The film elevated the Mustang's status as a symbol of coolness and rebellion, influencing car enthusiasts and boosting Mustang sales.

- **Legacy:** "Bullitt" remains a benchmark for car chases, inspiring countless films and securing the Mustang's place in automotive and film lore.

Chevrolet Camaro as "Bumblebee" in "Transformers" (2007)

The Chevrolet Camaro played a pivotal role in the "Transformers" franchise, embodying the character Bumblebee.

- **Character Representation:** As Bumblebee, the Camaro symbolizes loyalty, bravery, and transformation, aligning with the character's traits in the series.

- **Design and Technology:** The Camaro's aggressive styling and performance capabilities enhanced its appeal as a heroic and technologically advanced vehicle.

- **Brand Association:** The association with a beloved character reinforced Chevrolet's image as a maker of dynamic and high-performance cars, appealing to both younger audiences and longtime fans.

Other Notable Movie Appearances

- **"Gone in 60 Seconds" (2000):** Featured the Shelby GT500, emphasizing speed and the thrill of car theft, contributing to the mythos of the high-performance Mustang.

- **"Mad Max" Series:** While primarily featuring modified vehicles, the ruggedness and versatility of American trucks like the Ford F-150 influenced the design and perception of post-apocalyptic vehicles.

- **"Transformers" Series:** Beyond Bumblebee, other models like the Chevrolet Camaro ZL1 have appeared, further embedding Chevrolet's trucks and cars in popular media.

Influence in Music

Songs Celebrating Automobiles

Automobiles are a recurring theme in music, symbolizing freedom, status, and personal expression. Ford and Chevrolet vehicles frequently inspire song lyrics, music videos, and artist imagery.

- **"Mustang Sally" by Wilson Pickett:** Celebrates the allure and desirability of the Ford Mustang, embedding the car into the soul and rhythm of American music.

- **"Born to Run" by Bruce Springsteen:** References driving fast cars like the Mustang as metaphors for escaping and seeking a better life.

- **"Chevrolet" by ZZ Top:** Highlights the appeal of Chevrolet vehicles, associating them with a rock 'n' roll lifestyle and rebellious spirit.

Music Videos and Performances

Music videos often feature Ford and Chevrolet vehicles to enhance visual storytelling and connect with automotive enthusiasts.

- **Bon Jovi's "Living on a Prayer":** Features a Ford Mustang GT, reinforcing themes of resilience and striving for dreams.

- **Rage Against the Machine's "Renegades of Funk":** Showcases Chevrolet Camaro models, aligning the cars with themes of rebellion and counterculture.

Automotive Sponsorships and Collaborations

Ford and Chevrolet collaborate with musicians and bands, leveraging their brands in concert tours, music festivals, and artist endorsements.

- **Ford and Jay-Z:** Partnerships that blend automotive innovation with hip-hop culture, reaching diverse audiences.

- **Chevrolet and Metallica:** Sponsorships that associate Chevrolet vehicles with high-energy music genres, appealing to passionate and dynamic consumers.

Representation in Art

Automotive Artistry

Ford and Chevrolet vehicles inspire artists across various mediums, from painting and sculpture to digital art and photography.

- **Classic and Modern Depictions:** Artists capture both the timeless elegance of classic Mustangs and Camaros and the sleek innovation of modern electric models.

- **Symbolic Interpretations:** Vehicles are often portrayed as symbols of power, freedom, and technological prowess, reflecting societal values and aspirations.

- **Interactive and Digital Art:** With advancements in

technology, artists create interactive installations and digital artworks that celebrate automotive design and engineering.

Iconic Automotive Exhibits

Museums and galleries feature exhibitions dedicated to the history and design of Ford and Chevrolet vehicles, showcasing their impact on art and culture.

- **The Petersen Automotive Museum:** Hosts exhibits that highlight the artistic and cultural significance of iconic Ford and Chevrolet models.
- **Automotive Design Galleries:** Dedicated spaces where the design evolution of Mustangs, Camaros, F-Series, and Silverado trucks are explored through art and design showcases.

Collaborations with Artists

Ford and Chevrolet collaborate with renowned artists to create limited-edition models, artistic merchandise, and exclusive artwork that celebrate their vehicles.

- **Ford x Artist Collaborations:** Custom paint jobs and interior designs created in partnership with artists, turning each vehicle into a unique piece of art.
- **Chevrolet's Artistic Editions:** Special editions of the Camaro and Silverado featuring artwork inspired by music, film, and contemporary art movements.

Iconic Appearances and Enduring Symbols

Ford Mustang in "Bullitt"

The 1968 Ford Mustang GT 390 driven by Steve McQueen in

"Bullitt" remains one of the most iconic automotive moments in film history.

- **Cinematic Excellence:** The film's high-speed chases and the Mustang's standout performance showcased the car's capabilities, influencing both automotive enthusiasts and filmmakers.

- **Cultural Legacy:** The Mustang's appearance in "Bullitt" solidified its image as a symbol of coolness, performance, and American ingenuity.

- **Collectible Status:** The specific Mustang used in the film has become a highly sought-after collector's item, valued for its historical significance and cinematic heritage.

Chevrolet Camaro as "Bumblebee" in "Transformers"

The transformation of the Chevrolet Camaro into the beloved character Bumblebee in the "Transformers" franchise exemplifies the powerful synergy between automotive excellence and pop culture storytelling.

- **Character Enhancement:** The Camaro's design and performance capabilities were integral to Bumblebee's character, embodying agility, strength, and adaptability.

- **Brand Recognition:** Associating the Camaro with a popular and heroic character boosted Chevrolet's brand visibility and appeal among younger audiences.

- **Merchandising and Licensing:** The partnership led to extensive merchandise, including model cars, apparel, and collectibles, further embedding the Camaro in pop culture.

Other Noteworthy Appearances

- **Ford Bronco in "Jurassic Park" (1993):** The rugged Ford Bronco became synonymous with adventure and exploration, contributing to its iconic status.

- **Chevrolet Corvette in "The Fast and the Furious" Series:** The Corvette's high-performance image aligns with the series' themes of speed and adrenaline, enhancing its appeal among racing enthusiasts.

- **Ford F-150 in "Dumb and Dumber" (1994):** Showcased the F-150's reliability and ruggedness, reinforcing its image as a dependable workhorse.

Merchandise and Licensing

Model Cars and Collectibles

Ford and Chevrolet collaborate with manufacturers to produce scale models and collectibles of their iconic vehicles, catering to enthusiasts and collectors.

- **Scale Models:** Highly detailed replicas of Mustangs, Camaros, F-Series, and Silverado trucks available in various scales for display and collection.

- **Limited-Edition Collectibles:** Special editions featuring unique paint schemes, badges, and packaging that celebrate specific models or pop culture appearances.

Apparel and Accessories

Branded apparel and accessories allow fans to express their loyalty and passion for Ford and Chevrolet vehicles.

- **Clothing Lines:** T-shirts, hoodies, jackets, and hats

featuring iconic logos, vehicle designs, and thematic graphics related to specific models.

- **Accessories:** Items such as keychains, hats, bags, and automotive-themed jewelry that offer practical and stylish ways to showcase brand allegiance.

Video Games and Virtual Experiences

Ford and Chevrolet vehicles feature prominently in video games and virtual reality experiences, bridging the gap between real-world cars and digital entertainment.

- **Racing Games:** Presence in popular racing titles like "Forza Horizon" and "Need for Speed," allowing players to drive and customize their favorite Ford and Chevrolet models.

- **Virtual Showrooms:** Interactive virtual showrooms that offer immersive experiences of exploring and configuring vehicles in a digital environment.

Collaborative Merchandise

Joint collaborations with other brands and designers result in unique merchandise that appeals to a broader audience.

- **Ford x

Artist Collaborations:** Limited-edition apparel and accessories designed in partnership with contemporary artists, blending automotive design with artistic expression.

- **Chevrolet x Lifestyle Brands:** Collaborations with lifestyle and streetwear brands to create trendy, automotive-inspired clothing and accessories.

Fan Communities: Building Lifelong Loyalty

Car Clubs and Enthusiast Groups

Ford and Chevrolet support a multitude of car clubs and enthusiast groups that foster a sense of community and shared passion among owners and fans.

- **Ford Mustang Club of America (FMCA):** One of the largest and most active Mustang clubs, organizing events, rallies, and meet-ups that bring together Mustang enthusiasts from across the country.

- **Chevrolet Camaro Club of America (CCCA):** Provides a platform for Camaro owners to connect, share their experiences, and participate in exclusive events and activities.

- **Regional and Specialized Clubs:** Numerous regional clubs and specialized groups cater to specific models, interests, and customization preferences, enhancing community engagement and brand loyalty.

Online Forums and Discussion Boards

Digital platforms facilitate global connections among enthusiasts, enabling the exchange of ideas, tips, and shared experiences.

- **Mustang6G.com and Camaro5.com:** Dedicated forums where owners and fans discuss modifications, maintenance, and share their love for the Mustang and Camaro.

- **Reddit Communities:** Subreddits like r/Ford and r/Chevrolet provide spaces for broader discussions, news updates, and community-driven content.

- **Social Media Groups:** Facebook groups and Discord servers where members can interact, share photos, organize events, and support one another.

Events and Meet-Ups

Regular events and meet-ups organized by car clubs and the brands themselves create opportunities for face-to-face interactions and community building.

- **Annual Car Shows:** Events like the GM Car Show and Mustang Mania attract thousands of enthusiasts, showcasing a wide array of vehicles and fostering camaraderie.

- **Track Days and Rallies:** Organized track days and rallies allow owners to experience their vehicles' performance capabilities in a controlled environment, enhancing their connection to the brand.

- **Charity and Community Events:** Participation in charitable events and community service initiatives helps build positive brand associations and strengthens community ties.

Interactive Online Platforms

Engaging online platforms enhance the sense of community and provide continuous interaction opportunities.

- **Virtual Car Meets:** Online meet-ups and live streams where enthusiasts can showcase their vehicles, share stories, and participate in discussions regardless of geographical barriers.

- **Interactive Challenges and Competitions:** Hosting online challenges, such as photo contests or customization competitions, encourages active participation and creativity within the community.

- **Live Q&A Sessions:** Interactive sessions with designers, engineers, and brand ambassadors provide fans with insider insights and foster a deeper connection with the brands.

Influence on Brand Image and Consumer Perception

Reinforcing Brand Heritage

Pop culture appearances and community engagement reinforce the rich heritage of Ford and Chevrolet, celebrating their legacy while highlighting their ongoing evolution.

- **Mustang and Camaro as Cultural Symbols:** The enduring popularity of these models in media and fan communities underscores their status as cultural icons, embodying the spirit of American automotive excellence.

- **F-Series and Silverado as Workhorse Legends:** Representations in films and community activities highlight the reliability and strength of Ford's F-Series and Chevrolet's Silverado, reinforcing their reputations as dependable workhorses.

Modernizing Brand Image

Integrating modern technology and sustainable practices into classic models and pop culture representations helps modernize the brands' images, appealing to contemporary consumers.

- **Electric and Autonomous Innovations:** Showcasing electric models like the Mustang Mach-E and Silverado EV in media and community events emphasizes the brands' commitment to innovation and sustainability.

- **Advanced Safety and Connectivity:** Highlighting advanced safety features and connectivity options in media appearances and community engagements aligns the brands with modern consumer priorities.

Enhancing Emotional Connections

Pop culture representations and active community engagement create emotional connections between the brands and their consumers, fostering long-term loyalty.

- **Nostalgia and Modernity:** Balancing nostalgic elements with modern advancements in media portrayals and community events appeals to both longtime fans and new generations.

- **Shared Experiences:** Facilitating shared experiences through events, online interactions, and collaborative initiatives strengthens the emotional bond between the brands and their customers.

Legacy of Pop Culture Integration

Enduring Popularity and Influence

The integration of Ford and Chevrolet vehicles into pop culture has created a lasting legacy that continues to influence automotive trends and consumer preferences.

- **Timeless Appeal:** Iconic appearances in movies, music, and art ensure that Ford and Chevrolet remain relevant and admired across generations.

- **Inspirational Impact:** The presence of these vehicles in various media inspires future designers, engineers, and enthusiasts, perpetuating the brands' influence on automotive culture.

Cultural Preservation and Evolution

By embracing their roles in pop culture, Ford and Chevrolet contribute to the preservation and evolution of automotive heritage, ensuring that their legacies endure.

- **Documenting History:** Media portrayals and community narratives document the brands' histories, achievements, and cultural significance for future generations.

- **Adapting to Change:** Continuously evolving their representations in pop culture allows the brands to adapt to changing societal values and technological advancements, maintaining their relevance and appeal.

Global Recognition and Impact

Ford and Chevrolet's presence in international pop culture further extends their influence, enhancing global brand recognition and fostering cross-cultural connections.

- **Global Icon Status:** Iconic models like the Mustang and Camaro are recognized and celebrated worldwide, transcending cultural and geographical boundaries.

- **International Collaborations:** Participation in global media projects and international fan communities strengthens the brands' global presence and impact.

Conclusion: The Symbiotic Relationship Between Automotive Excellence and Pop Culture

The interplay between Ford, Chevrolet, and pop culture is a testament to the enduring appeal and cultural significance of these automotive giants. Through their strategic appearances in movies, music, and art, and by fostering vibrant fan communities, Ford and Chevrolet have not only enhanced their brand images but also solidified their places in the cultural zeitgeist. This symbiotic relationship has created a powerful feedback loop where cultural representations inspire brand innovations, and brand achievements, in turn, influence cultural narratives.

As Ford and Chevrolet continue to innovate with electric vehicles, autonomous technologies, and sustainable practices, their presence in pop culture will undoubtedly evolve, reflecting and shaping the future of automotive excellence. The legacy of their impact on pop culture underscores the importance of adaptability, creativity, and community engagement in sustaining long-term brand relevance and loyalty.

The journey of Ford and Chevrolet through the realms of pop culture illustrates how automotive brands can transcend their traditional roles, becoming integral parts of the stories, music, and art that define societies. As these brands continue to navigate the digital age and beyond, their ability to blend heritage with innovation will ensure their enduring influence on both the automotive industry and the broader cultural landscape.

Chapter 17: The Enthusiasts and the Aftermarket Scene

Automobiles are more than just modes of transportation; they are canvases for personal expression, symbols of identity, and embodiments of passion. Within this realm, car enthusiasts play a pivotal role, transforming their vehicles into unique masterpieces through modifications and personalization. Ford and Chevrolet, with their diverse and iconic lineups, have long been favorites among these passionate individuals. This chapter delves into the vibrant world of car enthusiasts who modify and personalize their Ford and Chevrolet vehicles, exploring the motivations behind these transformations, the diverse facets of the aftermarket industry, and the significant economic impact it holds.

The Car Enthusiast Culture

Passion and Personal Expression

For many, modifying a vehicle is an extension of personal identity and a manifestation of creativity. Enthusiasts are driven by a desire to:

- **Enhance Performance:** Upgrading engines, exhaust systems, and suspension components to improve speed, handling, and overall driving dynamics.

- **Aesthetic Customization:** Altering the exterior and interior design through paint jobs, body kits, decals, and custom upholstery to reflect individual tastes.

- **Technological Upgrades:** Incorporating advanced electronics, infotainment systems, and lighting solutions to modernize and personalize the driving experience.

- **Heritage Preservation:** Restoring classic models to

their former glory or maintaining their unique character while integrating modern enhancements.

Community and Camaraderie

The car enthusiast culture thrives on community and shared experiences. Enthusiasts often come together through:

- **Car Clubs:** Local and national clubs dedicated to specific models or brands, providing a platform for members to share their projects, participate in events, and foster friendships.

- **Online Forums and Social Media Groups:** Digital spaces where enthusiasts discuss modifications, share advice, showcase their vehicles, and organize meet-ups.

- **Events and Gatherings:** Organized events such as car shows, meet-ups, drag races, and track days that celebrate automotive passion and innovation.

The Aftermarket Industry: A Thriving Ecosystem

Definition and Scope

The aftermarket industry encompasses a wide range of products and services designed to modify, enhance, or maintain vehicles after their initial sale. This includes:

- **Performance Parts:** Engine upgrades, exhaust systems, turbochargers, suspension components, and braking systems aimed at boosting vehicle performance.

- **Aesthetic Parts:** Body kits, spoilers, custom paint jobs, decals, wheels, and lighting enhancements that alter the vehicle's appearance.

- **Interior Upgrades:** Custom upholstery, advanced infotainment systems, steering wheels, shift knobs, and ergonomic enhancements for improved comfort and functionality.

- **Technological Add-Ons:** GPS systems, advanced driver-assistance systems (ADAS), audio systems, and connectivity devices that modernize the vehicle's technology.

Key Players in the Aftermarket

The aftermarket scene is populated by a mix of large-scale manufacturers, specialized boutiques, and independent workshops. Prominent companies and brands include:

- **Cobb Tuning:** Known for performance parts and tuning solutions for various Ford and Chevrolet models.

- **Roush Performance:** Specializes in high-performance Ford vehicles, offering a range of upgrades from engines to suspension systems.

- **Hennessey Performance:** Offers extensive modifications for Chevrolet models, particularly trucks and high-performance cars.

- **MagnaFlow:** Renowned for exhaust systems that enhance both performance and sound quality.

Customization Trends

The aftermarket industry continually evolves, with trends reflecting both technological advancements and shifting consumer preferences:

- **Electric Vehicle (EV) Modifications:** As EVs like the

Ford Mustang Mach-E and Chevrolet Bolt EV gain popularity, the aftermarket adapts with custom charging solutions, performance enhancements, and aesthetic upgrades tailored to electric powertrains.

- **Smart Technology Integration:** Incorporating smart devices, enhanced infotainment systems, and connectivity options into vehicles to provide a seamless and modern driving experience.

- **Sustainable Modifications:** Increasing demand for eco-friendly modifications, such as lightweight materials, energy-efficient components, and sustainable manufacturing practices.

Economic Impact of the Aftermarket Industry

Market Size and Growth

The automotive aftermarket is a multi-billion-dollar industry with substantial growth potential. Key factors contributing to its expansion include:

- **Vehicle Longevity:** As vehicles remain on the road longer, the demand for maintenance, repairs, and upgrades increases.

- **Technological Advancements:** Continuous innovations in automotive technology drive demand for new aftermarket products and services.

- **Consumer Spending:** Enthusiasts' willingness to invest in their vehicles fuels economic activity within the aftermarket sector.

Job Creation and Economic Contributions

The aftermarket industry significantly contributes to the

economy by:

- **Employment Opportunities:** Creating jobs across various segments, including manufacturing, retail, installation, and consultancy services.

- **Small Business Support:** Empowering small businesses and independent workshops to thrive by offering specialized services and unique products.

- **Revenue Generation:** Driving substantial revenue through the sale of parts, accessories, and services, which in turn supports broader economic growth.

Innovation and Competition

The competitive landscape of the aftermarket industry fosters innovation, as companies strive to offer cutting-edge solutions that meet the evolving needs of enthusiasts. This competition drives:

- **Product Quality and Variety:** Continuous improvement in product quality and the introduction of diverse offerings to cater to different preferences and performance goals.

- **Pricing Strategies:** Competitive pricing ensures affordability and accessibility, making customization attainable for a broader audience.

- **Customer Service and Support:** Enhanced customer service and technical support build trust and loyalty among consumers, further stimulating economic activity.

Notable Aftermarket Projects and Custom Builds

Ford Mustang Custom Builds

The Ford Mustang has been a favorite among enthusiasts for its performance potential and iconic design. Notable custom builds include:

- **Shelby GT500 Builds:** High-performance builds that incorporate supercharged engines, racing suspension systems, and aerodynamic enhancements.

- **Restomods:** Restoration projects that blend classic Mustang aesthetics with modern technology and performance upgrades, maintaining the car's heritage while enhancing its capabilities.

- **Track-Focused Mustangs:** Builds optimized for track performance, featuring lightweight components, advanced braking systems, and precision handling upgrades.

Chevrolet Camaro Custom Builds

The Chevrolet Camaro's muscular design and powerful performance make it a popular choice for customization. Notable custom builds include:

- **Z/28 Performance Builds:** Enhancements that include upgraded LS engines, performance exhaust systems, and race-tuned suspensions to maximize track performance.

- **Street Aesthetic Builds:** Customizations focused on enhancing the Camaro's visual appeal through unique paint jobs, body kits, and interior modifications.

- **Off-Road Capable Camaros:** Builds that adapt the

Camaro for off-road adventures, incorporating lift kits, all-terrain tires, and reinforced suspensions.

Car Clubs, Forums, and Events: Pillars of the Aftermarket Community

Car Clubs: Fostering Community and Collaboration

Car clubs are essential for building a sense of community among enthusiasts. They provide a platform for sharing knowledge, showcasing vehicles, and participating in collective activities.

- **Ford Mustang Club of America (FMCA):** Offers members exclusive access to events, technical resources, and community forums, fostering a strong sense of camaraderie among Mustang owners.

- **Chevrolet Camaro Club of America (CCCA):** Organizes national and regional events, providing a space for Camaro enthusiasts to connect, share modifications, and celebrate their vehicles.

Online Forums and Discussion Boards: Virtual Gathering Spaces

Online forums and discussion boards play a crucial role in the aftermarket scene, enabling enthusiasts to connect regardless of geographical boundaries.

- **Mustang6G.com:** A dedicated forum for sixth-generation Mustang owners to discuss modifications, maintenance, and share their projects.

- **Camaro5.com:** An active community where Camaro owners and fans exchange ideas, seek advice, and showcase their customized vehicles.

- **Reddit Communities:** Subreddits like r/Ford and r/Chevrolet provide platforms for broader discussions, news updates, and community-driven content.

Events and Gatherings: Celebrating Automotive Passion

Events and gatherings are the lifeblood of the aftermarket community, providing opportunities for enthusiasts to come together, showcase their vehicles, and engage in shared activities.

- **SEMA Show (Specialty Equipment Market Association):** The premier automotive specialty products trade event, where aftermarket manufacturers, car clubs, and enthusiasts gather to showcase the latest innovations and custom builds.

- **Ford Mustang Weekends:** Organized events featuring car shows, cruises, track days, and social gatherings that celebrate the Mustang heritage.

- **Chevrolet Camaro Nationals:** National events that bring together Camaro owners for exhibitions, contests, and social activities.

The Role of Media and Influencers in the Aftermarket Scene

Automotive Magazines and Publications

Print and digital publications play a significant role in shaping the aftermarket scene by:

- **Showcasing Custom Builds:** Featuring detailed articles and photo spreads of customized Ford and Chevrolet vehicles, inspiring enthusiasts with new ideas and trends.

- **Product Reviews and Guides:** Providing in-depth reviews of aftermarket parts and accessories, along with installation guides and performance tips.

- **Event Coverage:** Reporting on major automotive events, races, and car shows, keeping the community informed and engaged.

Social Media Influencers and Content Creators

Social media influencers and content creators amplify the reach of aftermarket trends and foster a sense of community through:

- **YouTube Channels:** Channels like "Street Muscle Magazine" and "Chevy Camaro GT" offer tutorials, build chronicles, and reviews that educate and inspire enthusiasts.

- **Instagram and TikTok:** Visual platforms where influencers share high-quality photos and short videos of custom builds, modifications, and automotive lifestyle content.

- **Live Streaming:** Platforms like Twitch and YouTube Live host live build sessions, Q&A discussions, and virtual car meets, allowing real-time interaction between creators and their audiences.

Television Shows and Documentaries

Television shows and documentaries bring the aftermarket scene to a broader audience, highlighting the creativity and dedication of car enthusiasts.

- **"Pimp My Ride":** Showcased extreme customizations of everyday vehicles, popularizing the concept of vehicle personalization.

- **"Top Gear" and "The Grand Tour":** Featured Ford and Chevrolet vehicles in challenges, reviews

, and special segments, enhancing their global appeal and visibility.

- **Documentaries:** Productions like "Restoration Garage" and "Hoonigan Garage" delve into the intricacies of car restoration and high-performance modifications, providing in-depth looks at the aftermarket process.

Economic Significance of the Aftermarket Industry

Market Size and Growth Potential

The automotive aftermarket is a substantial segment of the automotive industry, contributing significantly to overall economic growth. Key factors driving its expansion include:

- **Increased Vehicle Lifespan:** As vehicles remain operational longer, the demand for maintenance, repairs, and upgrades grows.

- **Technological Advancements:** Innovations in automotive technology create opportunities for new aftermarket products and services.

- **Consumer Demand for Personalization:** Enthusiasts' desire to personalize their vehicles drives the market for custom parts and accessories.

Job Creation and Business Opportunities

The aftermarket industry supports a diverse range of jobs and business opportunities, including:

- **Retail and Distribution:** Car part retailers, both

brick-and-mortar and online, facilitate the distribution of aftermarket products.

- **Manufacturing and Production:** Specialized manufacturers produce performance parts, aesthetic enhancements, and custom components.

- **Installation and Service:** Independent workshops and service centers offer installation and customization services, creating employment opportunities.

- **Design and Engineering:** Professionals in design and engineering develop innovative aftermarket products that meet the evolving needs of enthusiasts.

Revenue Generation and Economic Impact

The aftermarket industry generates significant revenue through the sale of parts, accessories, and services. This revenue supports:

- **Local Economies:** Independent businesses and local workshops benefit from aftermarket sales, contributing to community economic health.

- **Innovation and Research:** Investment in product development and research drives continuous improvement and technological advancements.

- **Global Trade:** International demand for aftermarket products fosters global trade, expanding market reach and economic influence.

Sustainability and Environmental Considerations

The aftermarket industry also has implications for sustainability and environmental responsibility:

- **Recycling and Reuse:** Promoting the recycling and reuse of automotive parts reduces waste and conserves resources.

- **Eco-Friendly Modifications:** Increasing demand for sustainable modifications, such as energy-efficient components and lightweight materials, aligns with global environmental goals.

- **Regulatory Compliance:** Adhering to environmental regulations in the production and distribution of aftermarket products ensures responsible industry practices.

Case Studies: Iconic Aftermarket Builds

Ford Mustang Shelby GT500 Custom Build

The Shelby GT500 is a prime example of aftermarket customization, blending factory performance with bespoke enhancements.

- **Performance Upgrades:** Installation of a supercharged 5.2-liter V8 engine, upgraded exhaust systems, and enhanced suspension components.

- **Aesthetic Enhancements:** Custom body kits, forged wheels, and signature Shelby badging that elevate the Mustang's appearance.

- **Technological Integration:** Advanced infotainment systems and driver-assistance features tailored to the owner's preferences.

- **Impact:** The customized Shelby GT500 not only delivers unparalleled performance but also stands as a testament to the creativity and dedication of Ford enthusiasts.

Chevrolet Camaro Z/28 Drag Build

The Camaro Z/28 is renowned for its drag racing capabilities, often customized for maximum speed and performance.

- **Engine Modifications:** High-horsepower LS3 or LS7 engines, turbochargers, and nitrous oxide systems for explosive acceleration.

- **Performance Enhancements:** Upgraded transmission, reinforced drivetrain, and high-performance brakes to handle intense drag racing conditions.

- **Safety Features:** Installation of roll cages, racing seats, and harnesses to ensure driver safety during high-speed runs.

- **Impact:** The Camaro Z/28 drag build exemplifies Chevrolet's legacy in performance and motorsports, showcasing the brand's ability to deliver vehicles that excel in competitive racing environments.

Ford F-150 Raptor Off-Road Build

The Ford F-150 Raptor is a favorite among off-road enthusiasts, often customized to tackle extreme terrains.

- **Suspension Upgrades:** Enhanced off-road suspension systems, larger shocks, and increased ground clearance for improved performance on rough trails.

- **Aesthetic Customizations:** Aggressive body kits, lift kits, and all-terrain tires that enhance both functionality and appearance.

- **Utility Enhancements:** Installation of winches, roof

racks, and lighting systems to support off-road adventures.

- **Impact:** The customized F-150 Raptor demonstrates Ford's commitment to versatility and performance, catering to enthusiasts who seek both power and rugged capability in their vehicles.

The Future of the Aftermarket Industry

Technological Integration and Smart Modifications

As vehicles become increasingly connected and technologically advanced, the aftermarket industry is evolving to incorporate smart modifications:

- **Connected Devices:** Integration of smart devices and IoT (Internet of Things) technologies that enhance vehicle connectivity and functionality.

- **Advanced Diagnostics:** Aftermarket diagnostic tools and software that provide real-time data and performance analytics.

- **Customization Software:** Development of advanced customization platforms that allow enthusiasts to virtually design and preview modifications before implementation.

Sustainable and Eco-Friendly Practices

The future of the aftermarket industry emphasizes sustainability and environmental responsibility:

- **Eco-Friendly Parts:** Increased production of sustainable aftermarket parts made from recycled materials or designed to improve fuel efficiency and reduce emissions.

- **Green Manufacturing:** Adoption of environmentally friendly manufacturing processes that minimize waste and energy consumption.

- **Recycling Programs:** Expansion of recycling initiatives that encourage the reuse and repurposing of automotive components.

Autonomous and Electric Vehicle Modifications

With the rise of electric and autonomous vehicles, the aftermarket industry is adapting to meet new demands:

- **Electric Performance Upgrades:** Development of performance parts specifically designed for electric powertrains, such as enhanced battery systems and electric motors.

- **Autonomous Integration:** Custom modifications that support autonomous driving features, including advanced sensors and connectivity enhancements.

- **Charging Solutions:** Aftermarket charging stations and portable power solutions tailored for electric vehicles.

Global Expansion and Market Diversification

The aftermarket industry is poised for continued global expansion and diversification:

- **Emerging Markets:** Increased focus on emerging markets with growing automotive ownership and customization demand.

- **Cultural Adaptations:** Development of aftermarket products that cater to diverse cultural preferences and regional vehicle specifications.

- **Collaborative Innovations:** Partnerships between global and local aftermarket companies to drive innovation and meet specific market needs.

Conclusion: The Enduring Passion of Car Enthusiasts

The world of car enthusiasts and the aftermarket scene represents the heart and soul of the automotive industry. Ford and Chevrolet, with their iconic models and diverse lineups, provide the perfect canvas for enthusiasts to express their passion, creativity, and dedication. The aftermarket industry not only fuels this enthusiasm but also contributes significantly to economic growth, innovation, and cultural expression.

As technology continues to advance and consumer preferences evolve, the aftermarket industry will remain a dynamic and integral part of the automotive landscape. Ford and Chevrolet are well-positioned to lead the way, offering vehicles that inspire customization and fostering communities that celebrate automotive excellence. The enduring passion of car enthusiasts ensures that the legacy of Ford and Chevrolet will continue to thrive, driving the future of automotive innovation and personalization.

The synergy between manufacturers, aftermarket providers, and passionate enthusiasts creates a vibrant ecosystem that celebrates the freedom and creativity inherent in automotive culture. As we look to the future, the commitment of Ford and Chevrolet to support and engage with this community will be pivotal in maintaining their status as leaders in the automotive world, ensuring that the spirit of customization and personal expression remains as strong as ever.

Chapter 18: Case Studies—Notable Models and Their Stories

In the vast and competitive landscape of the automotive industry, certain models transcend their functional roles to become cultural icons, symbols of innovation, and benchmarks of performance. Ford and Chevrolet, two of America's most storied automakers, have produced a plethora of such models that not only defined their respective brands but also left indelible marks on the automotive world. This chapter delves into detailed case studies of these notable models, examining their development, impact, and the lessons learned that have influenced future designs and strategies.

1. Ford Mustang: The American Icon

Overview

Introduced in 1964, the Ford Mustang revolutionized the automotive market by creating the "pony car" segment—a class of American muscle cars characterized by affordable performance, sleek styling, and a long hood-short deck design. The Mustang's inception was a direct response to the growing demand for sporty, stylish vehicles that appealed to younger generations.

Impact

Market Disruption

The Mustang's launch was a watershed moment in automotive history. Its affordability, starting price of around $2,368, made it accessible to a broad audience. The car's success led to a surge in demand, resulting in the production of over one million Mustangs in the first two years alone.

Cultural Significance

The Mustang became synonymous with American freedom, youth culture, and rebellion. It featured prominently in films, music, and art, embedding itself deeply into the cultural fabric. Iconic appearances, such as Steve McQueen's Mustang in "Bullitt" (1968), cemented its status as a symbol of coolness and performance.

Brand Legacy

The Mustang established Ford as a leader in performance vehicles and set a standard for future models. Its enduring popularity has spanned over five decades, with each generation introducing innovations while preserving the core elements that define the Mustang's identity.

Lessons Learned

Balancing Heritage and Innovation

The Mustang's ability to evolve with the times while maintaining its heritage is a key factor in its sustained success. Ford has adeptly incorporated modern technologies, design enhancements, and performance upgrades without alienating long-time enthusiasts.

Market Responsiveness

Ford's swift response to market demands and trends demonstrated the importance of agility in product development. The creation of the pony car segment itself was a testament to Ford's ability to innovate and capture emerging consumer interests.

Embracing Cultural Integration

Leveraging the Mustang's presence in pop culture amplified

its appeal and relevance. Strategic marketing and collaborations with media franchises ensured that the Mustang remained a beloved icon across generations.

2. Chevrolet Camaro: The Rival Emerges

Overview

Launched in 1966 as a direct competitor to the Ford Mustang, the Chevrolet Camaro aimed to capture a share of the burgeoning pony car market. The Camaro offered similar styling and performance features but distinguished itself through its design nuances and performance capabilities.

Impact

Intense Brand Rivalry

The introduction of the Camaro intensified the rivalry between Ford and Chevrolet, leading to a fierce competition that pushed both brands to continuously innovate. This competition spurred advancements in performance, styling, and marketing strategies, benefiting consumers with improved offerings.

Performance Heritage

The Camaro quickly became known for its powerful engine options and superior handling. Models like the Z28 and SS variants garnered acclaim for their performance, making the Camaro a favorite among racing enthusiasts and muscle car aficionados.

Cultural Footprint

Similar to the Mustang, the Camaro made significant cultural inroads. Its portrayal as Bumblebee in the "Transformers" franchise revitalized the model's image, connecting it with a new generation of fans and enhancing its appeal in popular media.

Lessons Learned

Design Differentiation

While competing directly with the Mustang, Chevrolet ensured the Camaro had its distinct identity through unique design elements and performance characteristics. This differentiation was crucial in establishing the Camaro as a worthy rival and maintaining consumer interest.

Adaptability

The Camaro's multiple generations demonstrate Chevrolet's ability to adapt to changing market conditions and consumer preferences. Each redesign incorporated contemporary trends while retaining the essence of what makes the Camaro special.

Leveraging Media and Pop Culture

Chevrolet's strategic use of media appearances, such as in "Transformers," showcased the Camaro's versatility and performance, enhancing its brand image and attracting a diverse audience.

3. Ford F-150: Dominating the Truck Market

Overview

The Ford F-Series, particularly the F-150, has been a cornerstone of Ford's lineup since its introduction in 1948. The F-150 has evolved to become the best-selling truck in America and one of the best-selling vehicles overall, known for its reliability, performance, and versatility.

Impact

Market Leadership

The F-150's consistent sales dominance underscores its broad appeal across various consumer segments, including tradespeople, families, and off-road enthusiasts. Its ability to serve as both a workhorse and a comfortable family vehicle has been pivotal to its success.

Innovation and Technology

The F-150 has consistently integrated advanced technologies, from eco-friendly engine options like the PowerBoost hybrid to cutting-edge features like Pro Power Onboard, which offers onboard power for tools and appliances. These innovations have kept the F-150 relevant in an increasingly competitive market.

Cultural Significance

The F-150 is a symbol of American craftsmanship and ruggedness. Its presence in media, from action films to television shows, reinforces its image as a dependable and powerful vehicle.

Lessons Learned

Versatility and Broad Appeal

The F-150's ability to cater to a wide range of needs and preferences highlights the importance of versatility in vehicle design. By offering multiple configurations, engine options, and feature sets, Ford ensures the F-150 remains appealing to a diverse customer base.

Commitment to Innovation

Continuous investment in research and development has enabled Ford to incorporate the latest technologies into the F-150, enhancing performance, efficiency, and user experience. This commitment to innovation is essential for maintaining market leadership.

Customer-Centric Approach

Ford's focus on understanding and meeting customer needs, whether through durability for work purposes or comfort for family use, has been integral to the F-150's enduring popularity.

4. Chevrolet Silverado: The Benchmark Truck

Overview

Introduced in 1998, the Chevrolet Silverado quickly established itself as a formidable competitor in the full-size truck market. Known for its robust performance, durability, and advanced features, the Silverado has become synonymous with reliability and power.

Impact

Market Competitiveness

The Silverado has consistently ranked among the top-selling trucks in the United States, challenging Ford's F-150 with its own blend of performance, comfort, and technological advancements. This competition has driven both brands to elevate their offerings continuously.

Technological Advancements

Chevrolet has equipped the Silverado with state-of-the-art

technologies, including advanced towing systems, driver-assistance features, and connectivity options. Innovations like the Multi-Flex Tailgate and the Silverado Trailering Package have enhanced the truck's functionality and appeal.

Cultural Presence

The Silverado's depiction in media, such as action-packed movies and adventure-themed shows, has reinforced its image as a reliable and powerful vehicle capable of handling demanding tasks and thrilling escapades.

Lessons Learned

Emphasis on Technology and Features

By prioritizing the integration of advanced technologies and user-friendly features, Chevrolet has ensured the Silverado remains competitive and attractive to modern consumers seeking convenience and capability.

Building on Strengths

Chevrolet's focus on enhancing the Silverado's core strengths—power, durability, and versatility—has solidified its reputation as a benchmark in the truck market. This strategic focus ensures the Silverado meets and exceeds consumer expectations.

Responsive to Market Trends

The Silverado's evolution reflects Chevrolet's responsiveness to market trends, such as the increasing demand for eco-friendly options and enhanced connectivity. Adapting to these trends is crucial for maintaining relevance and appeal.

5. Ford Explorer: Pioneering the SUV Boom

Overview

The Ford Explorer, launched in 1991, played a significant role in popularizing the sport utility vehicle (SUV) segment in the United States. Combining off-road capability with family-friendly features, the Explorer became a versatile vehicle that appealed to a broad range of consumers.

Impact

Segment Leadership

The Explorer's success helped establish the SUV as a mainstream vehicle choice, leading to the proliferation of similar models across the industry. Its blend of utility, comfort, and performance set a standard for future SUVs.

Design and Performance

Over the years, the Explorer has undergone numerous redesigns to incorporate modern styling, advanced safety features, and improved performance capabilities. Innovations like four-wheel drive options and powerful engine choices have enhanced its off-road prowess and on-road performance.

Cultural Relevance

The Explorer's presence in popular media, including television shows and movies that emphasize adventure and family dynamics, has reinforced its image as a dependable and adventurous vehicle.

Lessons Learned

Capitalizing on Market Trends

Ford's foresight in recognizing and capitalizing on the growing demand for SUVs demonstrated the importance of aligning product offerings with emerging consumer preferences.

Continuous Evolution

The Explorer's ability to evolve with changing market demands—such as the shift towards more fuel-efficient engines and advanced safety features—has been key to its sustained success.

Versatility in Design

By offering a range of configurations, from two-row to three-row seating, and varying levels of trim and performance options, Ford has ensured the Explorer meets the diverse needs of its customer base.

6. Chevrolet Corvette: The Epitome of American Sports Cars

Overview

The Chevrolet Corvette, introduced in 1953, is an iconic American sports car renowned for its performance, design, and engineering excellence. Often referred to as "America's Sports Car," the Corvette has consistently pushed the boundaries of automotive performance and aesthetics.

Impact

Performance Milestones

The Corvette has set numerous performance milestones, from achieving high top speeds to incorporating advanced engineering features like lightweight materials and aerodynamic designs. Models like the Corvette C8, with its mid-engine layout, have redefined the

vehicle's performance capabilities and design paradigms.

Design Innovation

Chevrolet has continually evolved the Corvette's design, balancing classic elements with modern aesthetics. The distinctive styling, aggressive lines, and iconic features like the split rear window have made the Corvette a visual and engineering marvel.

Cultural Symbolism

The Corvette's representation in films, music, and art underscores its status as a symbol of speed, luxury, and American ingenuity. Its association with racing heritage and high-performance culture has solidified its place in automotive legend.

Lessons Learned

Embracing Bold Innovation

The Corvette's willingness to embrace bold innovations, such as the mid-engine layout in the C8 generation, demonstrates the importance of taking calculated risks to push the envelope in performance and design.

Maintaining Heritage

While innovating, Chevrolet has preserved the Corvette's heritage, ensuring that each generation retains the elements that make it uniquely Corvette. This balance between innovation and tradition is crucial for maintaining brand loyalty and appeal.

Marketing Excellence

Effective marketing strategies that highlight the Corvette's performance, design, and heritage have been instrumental in sustaining its iconic status. Chevrolet's ability to position the Corvette as a premier sports car has ensured its enduring popularity.

7. Ford Ranger: Reinventing the Compact Truck

Overview

The Ford Ranger, originally introduced in 1983, has undergone several revivals and redesigns to adapt to changing market demands. Positioned as a compact truck, the Ranger offers a balance of utility, performance, and maneuverability, catering to consumers seeking a versatile and manageable truck option.

Impact

Market Adaptation

The Ranger's reintroduction in the mid-2010s filled a niche in the compact truck segment, appealing to consumers who desired the functionality of a truck without the size and fuel consumption of full-size models.

Performance and Capability

Modern iterations of the Ranger feature robust engine options, advanced towing capacities, and off-road capabilities, making it a competitive choice in both urban and rugged environments.

Cultural Relevance

The Ranger's portrayal in media as a reliable and capable vehicle for both work and adventure has enhanced its appeal, positioning it as a practical yet aspirational choice for a diverse range of consumers.

Lessons Learned

Niche Targeting

Identifying and targeting specific market niches, such as the compact truck segment, allows automakers to address unmet consumer needs effectively. The Ranger's success underscores the value of strategic market positioning.

Versatile Design

The Ranger's design versatility, accommodating both urban commuting and off-road adventures, highlights the importance of offering adaptable features that cater to various consumer lifestyles.

Reinvention for Relevance

Reviving and reinventing existing models, like the Ranger, demonstrates the potential for legacy vehicles to find new life and relevance in evolving markets through thoughtful redesigns and feature enhancements.

8. Chevrolet Tahoe: The Full-Size SUV Leader

Overview

Introduced in 1995, the Chevrolet Tahoe has become a staple in the full-size SUV market, known for its spacious interior, powerful performance, and advanced features. The Tahoe caters to families, professionals, and off-road enthusiasts seeking a versatile and reliable vehicle.

Impact

Market Leadership

The Tahoe has consistently ranked among the top-selling full-size SUVs in the United States, competing closely with models like the Ford Expedition. Its combination of luxury, performance, and practicality has made it a preferred choice for a wide range of consumers.

Technological Integration

Chevrolet has equipped the Tahoe with cutting-edge technologies, including advanced infotainment systems, driver-assistance features, and robust towing capabilities, enhancing its functionality and appeal.

Cultural Representation

The Tahoe's presence in media, from action films to family-oriented shows, underscores its role as a dependable and versatile vehicle, reinforcing its image as a symbol of strength and reliability.

Lessons Learned

Balancing Luxury and Utility

Successfully balancing luxury features with practical utility ensures the Tahoe appeals to both comfort-seeking and performance-oriented consumers. This dual appeal is essential for maintaining market leadership in the competitive SUV segment.

Continuous Improvement

Ongoing enhancements in design, technology, and performance keep the Tahoe relevant and desirable. Chevrolet's commitment to refining the Tahoe based on consumer feedback and market trends exemplifies the importance of continuous improvement.

Targeted Marketing

Effective marketing strategies that highlight the Tahoe's versatility, technological advancements, and family-friendly features have been instrumental in sustaining its popularity and driving sales.

9. Ford Escape: Innovating the Compact SUV

Overview

The Ford Escape, launched in 2000, is a compact crossover SUV that combines the versatility of an SUV with the efficiency and maneuverability of a passenger car. The Escape caters to urban dwellers, small families, and adventure seekers looking for a practical and stylish vehicle.

Impact

Market Penetration

The Escape played a significant role in the rise of the compact SUV segment, appealing to consumers seeking a versatile vehicle that offers ample cargo space, fuel efficiency, and modern styling.

Technological Features

Ford has equipped the Escape with advanced technologies, including hybrid and plug-in hybrid powertrains, advanced driver-assistance systems, and connectivity features, enhancing its appeal to eco-conscious and tech-savvy consumers.

Cultural Relevance

The Escape's depiction in media as a reliable and versatile vehicle for both city commuting and outdoor adventures has reinforced its image as a practical yet stylish choice, appealing to a broad audience.

Lessons Learned

Embracing Hybrid Technology

Integrating hybrid and plug-in hybrid powertrains into the Escape lineup demonstrates the importance of embracing eco-friendly technologies to meet evolving consumer demands and regulatory requirements.

Versatile Design and Features

The Escape's design flexibility, offering multiple seating configurations and advanced features, highlights the importance of versatility in catering to diverse consumer needs and lifestyles.

Innovation and Adaptability

Ford's ability to adapt the Escape to incorporate the latest technologies and design trends ensures its continued relevance and competitiveness in the dynamic compact SUV market.

10. Chevrolet Silverado 1500: Evolving the Classic Truck

Overview

The Chevrolet Silverado 1500, part of the broader Silverado lineup, is a full-size pickup truck known for its robust performance, durability, and advanced features. The Silverado 1500 caters to both commercial users and personal truck enthusiasts, offering a blend of power, comfort, and versatility.

Impact

Market Leadership

The Silverado 1500 has consistently been a top contender in the full-size truck market, competing closely with Ford's F-150. Its reputation for reliability and performance has solidified its position as a market leader.

Technological Advancements

Chevrolet has integrated advanced technologies into the Silverado 1500, including trailering enhancements, advanced infotainment systems, and driver-assistance features, enhancing its functionality and user experience.

Cultural Significance

The Silverado 1500's presence in media, from rugged action films to family-oriented commercials, reinforces its image as a dependable and powerful vehicle, appealing to a wide range of consumers.

Lessons Learned

Innovation in Performance and Utility

Continuous innovation in performance and utility features ensures the Silverado 1500 remains competitive and appealing to consumers seeking both power and practicality in their trucks.

Customer-Centric Enhancements

Incorporating customer feedback into product development, such as enhancing towing capacities and improving interior comfort, demonstrates the importance of a customer-centric approach in sustaining product relevance and satisfaction.

Strategic Market Positioning

Positioning the Silverado 1500 as both a workhorse and a comfortable daily driver allows Chevrolet to capture a broad market segment, catering to diverse consumer needs and preferences.

Conclusion: Lessons from Notable Models

The case studies of Ford and Chevrolet's notable models reveal several key insights into what drives success in the automotive industry:

1. Innovation and Adaptation

Continuous innovation, whether in performance, technology, or design, is essential for maintaining competitiveness and meeting evolving consumer demands. Both Ford and Chevrolet have demonstrated the ability to adapt their models to incorporate the latest advancements while preserving the core attributes that define their brands.

2. Balancing Heritage and Modernity

Successful models strike a balance between honoring their heritage and embracing modernity. Preserving iconic design elements and brand legacy while integrating contemporary features ensures enduring appeal across generations.

3. Market Responsiveness

Being attuned to market trends and consumer preferences allows manufacturers to capitalize on emerging opportunities and address unmet needs. The ability to swiftly respond to shifts in demand, such as the rise of SUVs or electric vehicles, is crucial for sustained success.

4. Strong Brand Identity

Cultivating a strong and distinct brand identity helps models stand out in a crowded market. Clear brand positioning, whether as a performance icon or a versatile utility vehicle, reinforces consumer recognition and loyalty.

5. Community and Cultural Integration

Integrating models into cultural narratives through media, music, and art enhances their symbolic significance and emotional resonance. Building and nurturing communities around these models fosters deep brand loyalty and advocacy.

6. Versatility and Broad Appeal

Offering versatility in design, configurations, and features ensures models can cater to a wide range of consumer needs and preferences. This broad appeal is vital for capturing diverse market segments and driving sales growth.

7. Customer-Centric Approach

Focusing on customer needs and feedback in product development ensures that models remain relevant and desirable. A customer-centric approach fosters satisfaction, loyalty, and positive word-of-mouth, further enhancing market success.

8. Strategic Marketing and Media Presence

Effective marketing strategies that leverage media appearances, influencer partnerships, and community engagement amplify the visibility and desirability of models. Strategic media integration solidifies models' positions as cultural icons and symbols of excellence.

9. Sustainability and Technological Leadership

Embracing sustainability and leading in technological advancements position models for future success in an increasingly eco-conscious and tech-driven market. Investing in electric and autonomous technologies ensures readiness for the automotive industry's next evolution.

10. Economic Contributions and Aftermarket Support

Strong support from the aftermarket industry, including performance parts and customization options, enhances models' appeal and longevity. The economic contributions of the aftermarket scene also underscore its significance in driving innovation and consumer engagement.

Final Thoughts

The success stories of Ford and Chevrolet's notable models underscore the importance of a multifaceted approach to automotive excellence. By blending innovation, heritage, market responsiveness, and cultural integration, these models have not only achieved commercial success but also left lasting legacies that continue to influence the automotive landscape. The lessons learned from these case studies provide valuable insights for current and future automotive endeavors, highlighting the enduring principles that drive success in a dynamic and competitive industry.

As Ford and Chevrolet continue to evolve, the foundational strategies that propelled their iconic models to greatness will remain essential. Embracing change, fostering community, and maintaining a commitment to quality and innovation will ensure that these legendary brands continue to shape the future of automotive excellence.

Chapter 19: Challenges and Controversies

In the highly competitive and scrutinized automotive industry, even the most reputable manufacturers are not immune to challenges and controversies. Ford and Chevrolet, two of America's most storied automakers, have faced their share of recalls, legal battles, and public controversies over the decades. This chapter examines some of the most significant issues these companies have encountered, analyzes the strategies they employed to manage crises, and evaluates their recovery and resilience in the face of adversity.

Introduction

The journey of any major automotive manufacturer is often punctuated by moments of triumph and tribulation. For Ford and Chevrolet, navigating recalls, legal disputes, and public controversies has been integral to their evolution and operational strategies. These challenges not only test the companies' commitment to quality and safety but also influence their public perception, financial health, and long-term sustainability. Understanding how Ford and Chevrolet have addressed these issues provides valuable insights into effective crisis management, corporate responsibility, and brand resilience.

1. Recalls: Ensuring Safety and Quality

Recalls are critical in the automotive industry as they address safety and quality issues that could potentially harm consumers. Both Ford and Chevrolet have initiated numerous recalls to rectify defects and comply with regulatory standards.

Ford's Notable Recalls

Ford Explorer Tire Recall (2000)

- **Issue:** In 2000, Ford recalled approximately 4.5 million Explorer SUVs due to a combination of Firestone tires and Explorer designs that increased the risk of tire blowouts and rollovers.

- **Impact:** The recall highlighted issues related to tire performance and vehicle stability, leading to significant safety concerns and public scrutiny.

- **Resolution:** Ford worked closely with Firestone to replace defective tires and implemented design changes to enhance vehicle stability. The recall effort included extensive communication with affected customers and regulatory agencies.

Ford Focus Door Latch Recall (2019)

- **Issue:** In 2019, Ford recalled over 800,000 Focus models due to faulty door latches that could prevent doors from being securely closed, posing safety risks.

- **Impact:** The defect raised concerns about vehicle safety and reliability, prompting consumer complaints and negative media coverage.

- **Resolution:** Ford issued a recall notice to affected owners, offering free repairs to replace the faulty door latches. The company emphasized transparency and swift action to restore consumer trust.

Ford Fusion Transmission Recall (2021)

- **Issue:** In 2021, Ford recalled certain Fusion models due to potential transmission issues that could lead to unexpected shifting or stalling.

- **Impact:** Transmission problems affected vehicle performance and safety, leading to decreased consumer confidence in the Fusion model.

- **Resolution:** Ford provided software updates and transmission replacements where necessary. The company also enhanced quality control measures to prevent future occurrences.

Chevrolet's Notable Recalls

Chevrolet Cruze Ignition Switch Recall (2014)

- **Issue:** Although part of a broader General Motors (GM) recall, certain Chevrolet Cruze models were affected by faulty ignition switches that could unexpectedly move from the "run" position to the "accessory" or "off" positions.

- **Impact:** The defect was linked to several fatalities and injuries, raising severe safety concerns and leading to widespread criticism of GM's recall management.

- **Resolution:** GM, including Chevrolet, initiated a comprehensive recall campaign to replace defective ignition switches. The company faced significant legal and financial repercussions, emphasizing the critical need for prompt and effective recall processes.

Chevrolet Bolt EV Battery Recall (2021)

- **Issue:** In 2021, Chevrolet recalled approximately 68,000 Bolt EVs due to potential battery defects that could increase the risk of battery fires.

- **Impact:** The recall affected Chevrolet's reputation for producing reliable and safe electric vehicles, especially as the Bolt EV was a key model in their electric lineup.

- **Resolution:** Chevrolet offered battery replacements and software updates to address the defect. The company also worked to enhance battery safety protocols and engaged in transparent communication with affected owners.

Chevrolet Silverado Airbag Recall (2020)

- **Issue:** Part of the Takata airbag recall, certain Chevrolet Silverado models were recalled due to faulty airbags that could deploy improperly, causing injury or death.

- **Impact:** The recall impacted Chevrolet's full-size truck lineup, leading to concerns about vehicle safety and manufacturing quality.

- **Resolution:** Chevrolet facilitated the replacement of defective airbags through authorized dealers, ensuring compliance with safety regulations and reinforcing their commitment to consumer safety.

2. Legal Battles: Navigating the Courtroom

Legal battles in the automotive industry can stem from various issues, including safety defects, labor disputes, and intellectual property rights. Ford and Chevrolet have both been involved in significant legal cases that have shaped their operational and strategic approaches.

Ford's Notable Legal Battles

Ford Pinto Case (1970s)

- **Issue:** The Ford Pinto was involved in a highly publicized legal case due to its vulnerability to rear-end collisions, which could lead to fires. Allegations arose that Ford was aware of the defect but chose not to implement safety improvements to save costs.

- **Impact:** The case resulted in widespread media coverage, damaging Ford's reputation and leading to significant financial losses due to lawsuits and settlements.

- **Resolution:** Ford faced numerous lawsuits, culminating in a landmark case that emphasized corporate responsibility and the importance of prioritizing safety over cost savings. The negative publicity spurred Ford to enhance its safety standards and regulatory compliance practices.

Ford Motor Company v. United Auto Workers (2019)

- **Issue:** This legal battle involved labor negotiations and disputes over wages, benefits, and working conditions between Ford Motor Company and the United Auto Workers (UAW) union.

- **Impact:** The dispute had potential implications for production schedules, labor costs, and employee morale, affecting Ford's operational efficiency.

- **Resolution:** Through negotiations and mediation, Ford and the UAW reached a tentative agreement that addressed key concerns, ensuring continuity in production and maintaining positive labor relations.

Chevrolet's Notable Legal Battles

Chevrolet Malibu Emissions Scandal (2016)

- **Issue:** Certain Chevrolet Malibu models were found to have emissions control issues, leading to violations of environmental regulations. The company was accused of manipulating emissions data to meet regulatory standards.

- **Impact:** The scandal resulted in regulatory fines, legal penalties, and a tarnished reputation for Chevrolet's commitment to environmental responsibility.

- **Resolution:** Chevrolet implemented corrective measures, including software updates and emissions system repairs. The company also invested in more rigorous compliance and quality control processes to prevent future violations.

Chevrolet Camaro Trademark Dispute (2000)

- **Issue:** Chevrolet faced a trademark dispute over the Camaro name, as another company claimed prior rights to the name in certain international markets.

- **Impact:** The dispute threatened Chevrolet's ability to market the Camaro under its iconic name globally, potentially impacting sales and brand recognition.

- **Resolution:** Through legal negotiations and settlements, Chevrolet secured the rights to use the Camaro name in key markets, allowing them to continue marketing the model without interruption. The case highlighted the importance of robust intellectual property protection strategies.

Chevrolet Silverado Theft Prevention Lawsuit (2018)

- **Issue:** Owners of Chevrolet Silverado trucks filed lawsuits alleging that the vehicles lacked adequate theft prevention features, making them susceptible to break-ins and theft.

- **Impact:** The lawsuits raised concerns about vehicle security and prompted calls for enhanced anti-theft technologies in Chevrolet's truck lineup.

- **Resolution:** Chevrolet responded by integrating advanced security features, such as improved alarm systems and tracking technologies, into the Silverado models. The company also engaged in settlement agreements with affected customers to address their grievances.

3. Controversies: Public Perception and Brand Image

Controversies can significantly impact a company's public image, customer trust, and financial performance. Ford and Chevrolet have navigated various controversies that have tested their brand integrity and response strategies.

Ford's Notable Controversies

Sync Infotainment System Privacy Concerns (2017)

- **Issue:** The Ford Sync infotainment system faced scrutiny over privacy concerns related to data collection and potential unauthorized access to user information.

- **Impact:** Concerns about data privacy led to increased scrutiny from consumers and regulators, affecting Ford's reputation for technological innovation.

- **Resolution:** Ford enhanced its data security measures, provided clearer privacy policies, and offered software updates to address vulnerabilities. The company also engaged in public communication campaigns to reassure customers about the safety and privacy of their data.

Ford's Dieselgate Allegations (2015)

- **Issue:** Ford was implicated in allegations of emissions cheating similar to the Volkswagen Dieselgate scandal, where diesel engines were found to emit pollutants beyond regulatory limits.

- **Impact:** The allegations damaged Ford's reputation for environmental responsibility and led to potential regulatory fines and legal actions.

- **Resolution:** Ford conducted internal investigations, cooperated with regulatory authorities, and implemented stricter emissions controls across its diesel vehicle lineup. The company also accelerated its shift towards more sustainable and environmentally friendly technologies.

Chevrolet's Notable Controversies

Chevrolet Volt Battery Recall (2016)

- **Issue:** The Chevrolet Volt, an extended-range electric vehicle, faced a recall due to potential battery fires caused by faulty wiring.

- **Impact:** The recall raised safety concerns and impacted consumer confidence in Chevrolet's electric vehicle technology.

- **Resolution:** Chevrolet initiated a recall campaign to

inspect and replace faulty batteries, ensuring the safety of affected vehicles. The company also reinforced its commitment to electric vehicle safety and reliability through enhanced quality control measures.

Chevrolet Silverado Plow Attachment Controversy (2019)

- **Issue:** The Chevrolet Silverado's plow attachment system was found to be prone to detachment during use, posing safety hazards and operational challenges for users.

- **Impact:** The issue led to safety concerns among truck owners and negative media coverage, affecting the Silverado's reputation for durability and functionality.

- **Resolution:** Chevrolet redesigned the plow attachment system to enhance its stability and safety. The company offered free modifications to affected owners and provided clear instructions on proper attachment procedures, demonstrating responsiveness to customer feedback and safety concerns.

4. Crisis Management: Strategies and Recovery

Effective crisis management is crucial for automotive companies to mitigate the impact of recalls, legal battles, and controversies. Ford and Chevrolet have employed various strategies to address and recover from these challenges, emphasizing transparency, accountability, and customer-centric approaches.

Ford's Crisis Management Strategies

Transparency and Communication

- **Proactive Disclosure:** Ford has adopted a proactive approach in disclosing recalls and safety issues, ensuring timely communication with affected customers and regulatory bodies.

- **Public Relations Campaigns:** Following significant recalls or controversies, Ford launches PR campaigns to inform the public about corrective actions and reinforce its commitment to safety and quality.

Customer Support and Remediation

- **Recall Services:** Ford provides free repairs, replacements, and updates for recalled vehicles, prioritizing customer safety and satisfaction.

- **Compensation Programs:** In cases of severe defects or accidents, Ford offers compensation and support to affected customers, demonstrating accountability and empathy.

Internal Improvements and Preventative Measures

- **Enhanced Quality Control:** Ford invests in improving manufacturing processes and quality control measures to prevent future defects and recalls.

- **Employee Training:** Comprehensive training programs for employees ensure adherence to safety standards and regulatory compliance.

Legal and Regulatory Compliance

- **Cooperation with Authorities:** Ford maintains

cooperative relationships with regulatory agencies, ensuring compliance with safety and environmental standards.

- **Legal Settlements:** When necessary, Ford engages in legal settlements to resolve disputes and avoid prolonged litigation, balancing financial prudence with brand reputation.

Chevrolet's Crisis Management Strategies

Swift Response and Remediation

- **Immediate Action:** Chevrolet responds promptly to recalls and safety issues, initiating recall campaigns and providing necessary repairs or replacements without delay.

- **Customer Communication:** Clear and consistent communication with customers about the nature of the issue, steps being taken to resolve it, and how to participate in recall programs.

Strengthening Safety and Quality Protocols

- **Quality Enhancements:** Following controversies or recalls, Chevrolet implements enhanced safety and quality protocols to prevent recurrence of similar issues.

- **Investment in R&D:** Increased investment in research and development ensures that Chevrolet's vehicles meet higher safety and performance standards.

Public Relations and Brand Rebuilding

- **Rebuilding Trust:** Chevrolet undertakes PR

initiatives to rebuild trust with consumers, emphasizing lessons learned and improvements made post-crisis.

- **Positive Storytelling:** Highlighting positive aspects of their vehicles, technological advancements, and community engagement to shift focus away from negative incidents.

Legal Strategies and Settlements

- **Legal Compliance:** Chevrolet ensures strict adherence to legal requirements, cooperating with investigations and regulatory processes.

- **Strategic Settlements:** Engaging in fair settlements to resolve legal disputes efficiently, minimizing financial and reputational damage.

5. Comparative Analysis: Ford vs. Chevrolet in Crisis Management

Approach to Transparency

- **Ford:** Known for its proactive transparency, Ford often leads with early disclosure of issues, comprehensive communication strategies, and public acknowledgment of mistakes. This approach helps in maintaining consumer trust even during crises.

- **Chevrolet:** Chevrolet also emphasizes transparency but has occasionally faced criticism for delayed responses in certain situations. However, their efforts to enhance communication post-controversies have strengthened their crisis management framework.

Customer-Centric Remediation

- **Ford:** Ford prioritizes customer safety and satisfaction through extensive recall services, compensation programs, and continuous support initiatives. Their focus on empathy and accountability resonates well with consumers.

- **Chevrolet:** Chevrolet similarly prioritizes customer remediation but has sometimes been challenged by the scale and complexity of certain recalls. Their ability to swiftly implement corrective measures has improved over time, aligning closely with Ford's customer-centric approach.

Internal Improvements and Preventative Measures

- **Ford:** Ford invests significantly in quality control and employee training post-crises, ensuring systemic changes to prevent future issues. Their commitment to continuous improvement is evident in their operational strategies.

- **Chevrolet:** Chevrolet has ramped up its quality assurance processes and invested in advanced technologies to enhance vehicle safety and reliability. Lessons learned from past controversies have driven substantial internal reforms.

Public Relations and Brand Rebuilding

- **Ford:** Ford effectively utilizes public relations to communicate recovery efforts, emphasizing lessons learned and showcasing enhanced product features. Their storytelling often highlights resilience and dedication to excellence.

- **Chevrolet:** Chevrolet focuses on positive storytelling

and community engagement to rebuild brand image. Their PR campaigns often highlight technological innovations and sustainability initiatives to overshadow past controversies.

Legal and Regulatory Compliance

- **Ford:** Ford maintains a cooperative stance with regulatory bodies, ensuring compliance and facilitating smooth resolution of legal disputes. Their legal strategies are balanced to protect both financial interests and brand reputation.

- **Chevrolet:** Chevrolet adheres strictly to legal requirements and has developed robust legal strategies to manage disputes. Their approach emphasizes swift resolution and compliance to mitigate prolonged legal challenges.

6. Lessons Learned and Best Practices

Importance of Proactive Communication

- **Early Disclosure:** Addressing issues promptly prevents misinformation and reduces the spread of negative sentiments. Both Ford and Chevrolet have demonstrated the value of early and transparent communication in crisis situations.

Customer Trust and Accountability

- **Prioritizing Safety:** Ensuring customer safety above all else fosters trust and loyalty. Demonstrating accountability through remediation efforts reassures consumers of the brands' commitment to quality.

Continuous Improvement and Innovation

- **Learning from Mistakes:** Analyzing and learning from past challenges helps prevent recurrence. Implementing systemic changes and investing in quality control are essential for long-term sustainability.

Robust Crisis Management Frameworks

- **Preparedness:** Having established crisis management protocols enables swift and effective responses. Both companies benefit from structured frameworks that guide their actions during emergencies.

Balancing Legal and Public Relations Strategies

- **Integrated Approach:** Coordinating legal strategies with public relations efforts ensures coherent and unified responses. This balance is crucial for maintaining brand integrity and minimizing reputational damage.

Embracing Technological Solutions

- **Advanced Monitoring:** Utilizing advanced technologies for monitoring vehicle performance and safety can preempt potential issues. Investments in R&D and innovative solutions enhance overall vehicle reliability.

Engaging with Stakeholders

- **Collaborative Efforts:** Engaging with stakeholders, including customers, regulators, and industry partners, fosters collaborative solutions and reinforces positive relationships.

Conclusion: Resilience and Recovery

The challenges and controversies faced by Ford and Chevrolet underscore the complexities of operating in the automotive industry. Recalls, legal battles, and public controversies test the resilience and adaptability of these iconic brands. Through proactive communication, customer-centric remediation, internal reforms, and strategic crisis management, both Ford and Chevrolet have demonstrated their ability to navigate adversity and emerge stronger.

These experiences highlight the critical importance of transparency, accountability, and continuous improvement in sustaining brand integrity and consumer trust. As the automotive landscape continues to evolve with technological advancements and shifting consumer expectations, the lessons learned from past challenges will inform Ford and Chevrolet's strategies for future success.

Ultimately, the ability to effectively manage crises and recover from setbacks not only safeguards the brands' reputations but also reinforces their positions as leaders in the global automotive market. Ford and Chevrolet's commitment to excellence, innovation, and customer satisfaction ensures their enduring legacy and continued influence in the industry.

Chapter 20: Looking Ahead—The Future of the Rivalry

As the automotive industry stands on the brink of unprecedented transformation, the longstanding rivalry between Ford and Chevrolet is poised to evolve in response to emerging technologies, shifting market dynamics, and evolving consumer preferences. This chapter speculates on the future trajectory of these two automotive giants, exploring potential advancements, strategic initiatives, and market shifts that will shape their competition in the coming decades. By examining trends in electric vehicles (EVs), autonomous driving, connectivity, sustainability, and global expansion, we can anticipate how Ford and Chevrolet will navigate the challenges and opportunities that lie ahead.

1. The Electric Vehicle Revolution

Ford's Electric Ambitions

Ford has made significant strides in electrifying its lineup, signaling a strong commitment to electric mobility. Key initiatives include:

Mustang Mach-E and F-150 Lightning

- **Mustang Mach-E:** As one of Ford's flagship electric SUVs, the Mach-E has set the standard for performance and range in the EV segment, appealing to both Mustang enthusiasts and new EV consumers.

- **F-150 Lightning:** The electric version of Ford's best-selling truck, the F-150 Lightning, combines the utility and performance of the traditional F-150 with the benefits of electric propulsion, targeting both commercial and personal truck markets.

Expanding EV Portfolio

Ford plans to expand its EV offerings with models across various segments, including sedans, SUVs, and commercial vehicles. This diversification aims to capture a broader market share and meet the diverse needs of consumers transitioning to electric mobility.

Investment in Battery Technology

Ford is investing heavily in battery technology, including the development of its own battery cells and partnerships with leading battery manufacturers. Enhancing battery efficiency, reducing costs, and improving sustainability are central to Ford's EV strategy.

Chevrolet's Electric Strategy

Chevrolet is equally committed to electrification, with a robust strategy to electrify its lineup and lead in the EV market.

Chevrolet Bolt EV and Bolt EUV

- **Bolt EV and Bolt EUV:** These models have positioned Chevrolet as a key player in the affordable EV market, offering practical range and advanced features that appeal to a wide range of consumers.

Silverado EV and Blazer EV

- **Silverado EV:** The all-electric Silverado aims to compete directly with Ford's F-150 Lightning, offering impressive range, towing capacity, and performance.

- **Blazer EV:** Combining sporty design with electric efficiency, the Blazer EV caters to consumers seeking a stylish and environmentally friendly SUV.

Investment in Charging Infrastructure

Chevrolet is investing in expanding charging infrastructure partnerships and offering incentives for home charging solutions. Enhancing accessibility to charging stations is crucial for accelerating EV adoption and ensuring customer convenience.

Impact on the Market

The aggressive push towards electrification by both Ford and Chevrolet is reshaping the automotive market. Increased competition in the EV segment is driving innovation, reducing costs, and expanding consumer choices. As regulatory pressures for lower emissions intensify, Ford and Chevrolet's investments in EV technology position them favorably to meet future environmental standards and consumer demands.

2. Autonomous Driving and Advanced Driver Assistance Systems (ADAS)

Ford's Autonomous Initiatives

Ford BlueCruise and Autonomous Vehicles

- **BlueCruise:** Ford's hands-free driving technology, BlueCruise, allows for semi-autonomous driving on pre-mapped highways. Continuous enhancements are expected to expand its capabilities and coverage.

- **Partnerships with Tech Firms:** Collaborations with companies like Argo AI underscore Ford's commitment to developing fully autonomous vehicles (Level 4 and Level 5). These partnerships focus on advancing sensor technology, artificial intelligence, and real-time data processing.

Integration with Mobility Services

Ford is exploring the integration of autonomous technology with mobility services, such as ride-sharing and delivery platforms. This diversification aims to capitalize on new revenue streams and adapt to changing transportation paradigms.

Chevrolet's Autonomous Vision

Super Cruise and Autonomous Features

- **Super Cruise:** Chevrolet's advanced driver-assistance system, Super Cruise, offers hands-free driving on compatible highways, similar to Ford's BlueCruise. Continuous upgrades aim to enhance its functionality and user experience.

- **Collaborations with Autonomous Leaders:** Chevrolet is partnering with autonomous technology leaders to accelerate the development of self-driving capabilities, focusing on urban and suburban environments.

Autonomous Commercial Vehicles

Chevrolet is investing in autonomous commercial vehicles, targeting sectors like logistics, delivery, and fleet management. These vehicles promise increased efficiency, reduced operational costs, and enhanced safety for commercial operations.

Market Implications

The race towards autonomous driving technology is intensifying, with Ford and Chevrolet striving to lead in this transformative field. Success in developing reliable and safe autonomous vehicles will not only redefine their product offerings but also establish them as pioneers in the next era of transportation. The integration of ADAS and autonomous features enhances vehicle safety and user experience, driving consumer preference towards technologically advanced models.

3. Connectivity and Smart Vehicle Technologies

Ford's Connectivity Enhancements

FordPass and SYNC Innovations

- **FordPass:** An integrated mobile app that offers vehicle management tools, including remote start, vehicle tracking, and maintenance scheduling. Future enhancements will focus on deeper integration with smart home devices and personalized user experiences.

- **SYNC System Upgrades:** Continuous upgrades to the SYNC infotainment system incorporate advanced voice recognition, seamless smartphone integration, and enhanced navigation features.

Over-the-Air (OTA) Updates

Ford is expanding its use of OTA updates, allowing vehicles to receive software upgrades and new features without visiting a dealership. This capability ensures that Ford vehicles remain up-to-date with the latest technologies and improvements.

Chevrolet's Connectivity Strategy

Chevrolet MyLink and Infotainment Systems

- **MyLink:** Chevrolet's infotainment system offers intuitive controls, voice commands, and connectivity features that enhance the driving experience. Future iterations will include augmented reality displays and enhanced AI-driven personalization.

- **Advanced Infotainment Features:** Incorporating larger touchscreens, integrated streaming services, and enhanced connectivity options to meet the demands of tech-savvy consumers.

Connected Services and Telematics

Chevrolet is expanding its connected services offerings, including real-time vehicle diagnostics, predictive maintenance, and enhanced security features. Telematics solutions provide valuable data for improving vehicle performance and user satisfaction.

Impact on Consumer Experience

Enhanced connectivity and smart vehicle technologies are pivotal in shaping the future driving experience. Ford and Chevrolet's investments in connectivity not only improve vehicle functionality and user convenience but also create new opportunities for customer engagement and personalized services. As vehicles become more integrated with digital ecosystems, the distinction between automotive and technology industries will blur, emphasizing the importance of innovation in both fields.

4. Sustainability and Environmental Responsibility

Ford's Sustainability Initiatives

Carbon Neutrality Goals

Ford has committed to achieving carbon neutrality globally by 2050, with interim targets to reduce emissions across its operations. Key strategies include:

- **Renewable Energy Adoption:** Transitioning manufacturing facilities to 100% renewable energy sources, such as solar and wind power.

- **Energy Efficiency Improvements:** Implementing energy-efficient technologies and processes to minimize energy consumption.

- **Sustainable Materials:** Increasing the use of recycled and sustainable materials in vehicle production, including recycled plastics, bio-based textiles, and eco-friendly coatings.

Circular Economy Practices

Ford embraces the principles of the circular economy to reduce waste and promote the reuse and recycling of materials.

- **Vehicle Recycling Programs:** Ensuring end-of-life vehicles are dismantled responsibly, with materials recycled or repurposed.

- **Modular Design:** Designing vehicles with modular components that facilitate easier repairs, upgrades, and recycling.

Chevrolet's Environmental Commitments

General Motors' Sustainability Goals

As part of General Motors (GM), Chevrolet aligns with the company's ambitious sustainability targets aimed at reducing environmental impact.

- **Carbon Neutrality by 2040:** GM aims to achieve carbon neutrality across its global operations by 2040, with Chevrolet contributing through electric vehicle advancements and sustainable manufacturing practices.

- **Zero Waste to Landfill:** Achieving zero waste to landfill in all manufacturing facilities, emphasizing waste reduction and recycling.

Sustainable Manufacturing Processes

Chevrolet integrates sustainable practices into its manufacturing processes to minimize environmental footprint.

- **Green Manufacturing Technologies:** Implementing energy-efficient production lines, reducing water usage, and utilizing eco-friendly materials.

- **Sustainable Supply Chain Management:** Ensuring suppliers adhere to environmental standards and promoting the use of sustainable materials and practices.

Market and Regulatory Impact

Sustainability is becoming a critical factor in consumer decision-making and regulatory compliance. Ford and Chevrolet's proactive approach to environmental responsibility not only aligns with global sustainability trends but also positions them favorably in markets with stringent emissions and environmental regulations. These initiatives enhance brand reputation, attract eco-conscious consumers, and ensure long-term viability in an increasingly sustainability-focused market.

5. Global Expansion and Market Diversification

Ford's Global Strategies

Emerging Markets Focus

Ford is intensifying its focus on emerging markets, recognizing the growth potential in regions like Asia-Pacific, Latin America, and Africa.

- **Localized Product Offerings:** Developing vehicles tailored to the specific needs and preferences of consumers in different regions, such as compact cars for urban markets and rugged trucks for emerging economies.

- **Strategic Partnerships:** Forming joint ventures and partnerships with local manufacturers to enhance market penetration and leverage regional expertise.

Electric Mobility in Global Markets

Ford is expanding its electric vehicle offerings globally, adapting to regional regulations and consumer demands.

- **Localized EV Production:** Establishing manufacturing facilities in key regions to produce EVs closer to target markets, reducing costs and enhancing supply chain efficiency.

- **Regional Charging Infrastructure Development:** Collaborating with local governments and private partners to expand EV charging infrastructure, facilitating broader adoption of electric vehicles.

Chevrolet's Global Initiatives

Strengthening Presence in Key Markets

Chevrolet is bolstering its presence in

key global markets, including China, South America, and Europe.

- **China:** Leveraging joint ventures with local manufacturers to produce and distribute vehicles tailored to the Chinese market. The Chevrolet Onix and Cruze have been particularly successful, catering to the preferences of Chinese consumers.

- **South America:** Continuing strong sales in Brazil and Argentina with models like the Chevrolet Onix and Silverado, which have been adapted to meet regional demands.

- **Europe:** Expanding the Chevrolet brand in Europe with a focus on electric and hybrid models, responding to the region's stringent emissions standards and growing demand for sustainable vehicles.

Electric and Hybrid Vehicle Expansion

Chevrolet is accelerating its electrification strategy in global markets, introducing a range of electric and hybrid vehicles to meet diverse consumer needs.

- **Electric SUVs and Trucks:** Introducing models like the Chevrolet Blazer EV and Silverado EV in international markets, appealing to consumers seeking high-performance and eco-friendly vehicles.

- **Hybrid Powertrains:** Expanding hybrid options

across various models to provide efficient and flexible powertrain solutions for different market segments.

Market Diversification Strategies

Ford and Chevrolet are diversifying their product portfolios to cater to a wide range of consumer preferences and regional requirements.

- **Product Line Diversification:** Introducing a variety of vehicle types, including compact cars, SUVs, trucks, and commercial vehicles, to capture different market segments and enhance global competitiveness.

- **Adaptation to Local Preferences:** Customizing vehicle features, designs, and powertrains to align with local tastes, driving conditions, and regulatory requirements.

Impact on Global Competitiveness

Global expansion and market diversification enable Ford and Chevrolet to tap into new revenue streams, mitigate risks associated with market saturation in their home country, and enhance their global brand presence. By adapting to regional needs and investing in localized production and infrastructure, both brands strengthen their competitiveness in the global automotive arena.

6. Sustainability and Corporate Social Responsibility (CSR)

Ford's Sustainability and CSR Initiatives

Environmental Stewardship

Ford is committed to minimizing its environmental impact through comprehensive sustainability initiatives.

- **Eco-Friendly Manufacturing:** Implementing energy-efficient technologies, reducing waste, and utilizing sustainable materials in manufacturing processes.

- **Carbon Offset Programs:** Investing in projects that offset carbon emissions, such as reforestation and renewable energy initiatives.

Social Responsibility Programs

Ford engages in various CSR activities aimed at supporting communities and fostering social well-being.

- **Community Engagement:** Supporting education, workforce development, and community infrastructure projects in regions where Ford operates.

- **Diversity and Inclusion:** Promoting diversity and inclusion within the workforce, ensuring equal opportunities and fostering a culture of respect and collaboration.

Chevrolet's Sustainability and CSR Efforts

Environmental Initiatives

Chevrolet, under the umbrella of General Motors, prioritizes sustainability through a range of environmental initiatives.

- **Sustainable Manufacturing:** Adopting green manufacturing practices, reducing carbon emissions, and enhancing energy efficiency across all production facilities.

- **Recycling and Waste Reduction:** Implementing comprehensive recycling programs and waste reduction strategies to minimize environmental impact.

Corporate Social Responsibility

Chevrolet's CSR efforts focus on making positive contributions to society and supporting sustainable development.

- **Community Support:** Investing in programs that support education, healthcare, and economic development in local communities.

- **Sustainable Mobility Solutions:** Promoting sustainable mobility through the development of electric and hybrid vehicles, encouraging eco-friendly transportation options.

Long-Term Benefits

Commitment to sustainability and CSR not only enhances Ford and Chevrolet's reputations but also contributes to long-term business sustainability. These initiatives attract environmentally conscious consumers, comply with regulatory requirements, and foster goodwill in communities, ensuring enduring brand loyalty and market relevance.

7. Technological Advancements and Future Innovations

Ford's Technological Innovations

Advanced Manufacturing Technologies

Ford is embracing advanced manufacturing technologies to enhance production efficiency and vehicle quality.

- **Automation and Robotics:** Implementing automated

assembly lines and robotics to streamline production processes, reduce costs, and improve precision.

- **Additive Manufacturing:** Utilizing 3D printing for prototyping and producing complex vehicle components, accelerating development timelines and reducing material waste.

Artificial Intelligence and Data Analytics

Ford leverages AI and data analytics to optimize vehicle performance, enhance customer experiences, and drive innovation.

- **Predictive Maintenance:** Using data analytics to predict and address vehicle maintenance needs proactively, improving reliability and customer satisfaction.

- **Personalized Driving Experiences:** Incorporating AI-driven personalization features that adapt vehicle settings and preferences based on individual driver behavior and preferences.

Chevrolet's Future Technological Innovations

Electric and Autonomous Technologies

Chevrolet is at the forefront of integrating electric and autonomous technologies into its vehicle lineup.

- **Battery Innovations:** Developing high-capacity, long-lasting batteries that enhance the range and performance of electric vehicles.

- **Autonomous Features:** Continuously improving Super Cruise and exploring fully autonomous driving capabilities to stay competitive in the evolving market.

Connected Vehicle Ecosystems

Chevrolet is expanding its connected vehicle ecosystems to offer seamless integration with digital lifestyles.

- **Vehicle-to-Everything (V2X) Communication:** Implementing V2X technologies that enable vehicles to communicate with each other and with infrastructure, enhancing safety and traffic management.

- **Smart Home Integration:** Developing features that allow vehicles to interact with smart home devices, providing enhanced convenience and connectivity for users.

Impact on Future Mobility

Technological advancements by Ford and Chevrolet are shaping the future of mobility, emphasizing sustainability, safety, and connectivity. These innovations not only improve vehicle performance and user experience but also contribute to the development of smarter, more efficient transportation systems that address global challenges such as urban congestion and environmental sustainability.

8. Strategic Partnerships and Collaborations

Ford's Strategic Alliances

Technology Partnerships

Ford collaborates with leading technology companies to drive innovation and integrate advanced features into its vehicles.

- **Microsoft Collaboration:** Partnering with Microsoft to enhance Ford's connectivity and cloud computing capabilities, enabling more sophisticated data analytics and connected services.

- **Autonomous Technology Firms:** Continuing partnerships with firms like Argo AI to accelerate the development of autonomous driving systems.

Sustainability Collaborations

Ford engages in partnerships that support its sustainability goals and environmental initiatives.

- **Renewable Energy Partnerships:** Collaborating with renewable energy providers to ensure the use of clean energy in manufacturing and operations.

- **Recycling and Circular Economy Partners:** Partnering with organizations that promote recycling and the reuse of materials, advancing Ford's circular economy initiatives.

Chevrolet's Collaborative Efforts

Tech Industry Collaborations

Chevrolet forms alliances with technology leaders to integrate cutting-edge technologies into its vehicles.

- **Google Partnership:** Collaborating with Google to enhance in-car connectivity, infotainment systems, and voice recognition capabilities.

- **Autonomous Driving Collaborations:** Partnering with autonomous technology companies to develop and implement advanced self-driving features in Chevrolet vehicles.

Environmental and Social Partnerships

Chevrolet engages in collaborations that advance its environmental and social responsibility goals.

- **Environmental NGOs:** Working with environmental organizations to promote sustainability and implement eco-friendly practices in manufacturing and operations.

- **Community Organizations:** Partnering with local community groups to support social initiatives and enhance community well-being.

Benefits of Strategic Partnerships

Strategic partnerships enable Ford and Chevrolet to leverage external expertise, accelerate innovation, and enhance their competitive positioning. Collaborations with technology and sustainability leaders facilitate the integration of advanced features and sustainable practices, ensuring that both brands remain at the forefront of industry advancements and consumer expectations.

9. Market Shifts and Consumer Preferences

Changing Consumer Demands

Shift Towards Sustainability

Consumers are increasingly prioritizing sustainability in their purchasing decisions, favoring brands that demonstrate environmental responsibility and offer eco-friendly products.

- **Preference for EVs:** Growing demand for electric and hybrid vehicles is driving Ford and Chevrolet to expand their electrified lineups.

- **Sustainable Materials:** Consumers are seeking vehicles made with sustainable and recycled materials, influencing design and manufacturing practices.

Desire for Advanced Technology

Modern consumers expect vehicles to be equipped with the latest technologies that enhance safety, connectivity, and convenience.

- **Smart Features:** Integration of AI, advanced driver-assistance systems, and seamless connectivity with digital devices is becoming a standard expectation.

- **Customization and Personalization:** Consumers desire personalized driving experiences and the ability to customize vehicle features to suit their preferences.

Urbanization and Mobility Trends

Rise of Urban Mobility Solutions

Urbanization is reshaping transportation needs, emphasizing compact, efficient, and connected vehicles that are well-suited for city environments.

- **Compact SUVs and Crossovers:** Increased demand for smaller, more maneuverable vehicles that offer the versatility of SUVs without the bulk.

- **Shared Mobility Services:** Growth of ride-sharing and car-sharing services influences vehicle design and features, prioritizing connectivity and flexibility.

Integration with Smart Cities

As cities become smarter, vehicles are increasingly integrated into interconnected urban ecosystems, enhancing mobility and reducing congestion.

- **V2X Communication:** Vehicles communicate with infrastructure and other vehicles to optimize traffic flow and improve safety.

- **Autonomous Fleets:** Development of autonomous vehicle fleets for public transportation and delivery services is becoming more feasible, driven by advancements in autonomous technology.

Impact on Ford and Chevrolet's Strategies

To align with these market shifts and evolving consumer preferences, Ford and Chevrolet are adapting their strategies to offer vehicles that meet the demands of sustainability, advanced technology, and urban mobility. This involves continuous innovation, strategic product development, and a focus on customer-centric design to ensure that their offerings remain relevant and competitive in a rapidly changing market landscape.

10. Potential Market Shifts and Competitive Dynamics

Emergence of New Competitors

The automotive landscape is witnessing the rise of new competitors, particularly from the technology and EV sectors. Companies like Tesla, Rivian, and Lucid Motors are disrupting traditional market dynamics, challenging Ford and Chevrolet to innovate and differentiate their offerings.

Tesla's Influence

Tesla's success in the EV market has set new standards for performance, range, and technology, pushing established automakers to accelerate their electrification efforts and enhance their technological capabilities.

Tech Giants Entering Automotive

Companies like Apple and Google are exploring automotive ventures, particularly in autonomous driving and connected vehicle technologies, intensifying competition and fostering a more technology-driven automotive industry.

Shifts in Global Trade and Supply Chains

Global trade dynamics and supply chain resilience are becoming critical factors in the automotive industry, influenced by geopolitical tensions, trade agreements, and the need for sustainable sourcing.

Localization of Production

To mitigate risks associated with global supply chain disruptions, Ford and Chevrolet are focusing on localizing production in key markets, reducing dependency on international supply chains and enhancing operational flexibility.

Sustainable Sourcing Practices

Emphasizing sustainable and ethical sourcing of materials is becoming increasingly important, driven by consumer expectations and regulatory requirements. Ford and Chevrolet are investing in transparent supply chains and sustainable material sourcing to align with global sustainability standards.

Technological Convergence and Innovation

The convergence of automotive and technology industries is driving innovation in areas such as artificial intelligence, connectivity, and smart manufacturing.

AI and Machine Learning

AI and machine learning are enhancing vehicle performance, safety, and user experiences, enabling features like predictive maintenance, personalized driving settings, and advanced driver assistance systems.

Smart Manufacturing

Adoption of smart manufacturing technologies, including the Internet of Things (IoT), automation, and data analytics, is improving production efficiency, quality control, and operational agility for Ford and Chevrolet.

Impact on Rivalry Dynamics

The evolving competitive landscape necessitates that Ford and Chevrolet continuously innovate and adapt to maintain their market positions. The rivalry will likely intensify as both brands strive to lead in electrification, autonomous driving, and connectivity, while also addressing sustainability and global market challenges. Strategic differentiation, customer-centric innovation, and robust crisis management will be key to sustaining their competitive edge.

11. Strategic Initiatives for Future Success

Ford's Future Strategic Initiatives

Expansion of Electric and Autonomous Fleet

Ford aims to significantly expand its electric and autonomous vehicle offerings, targeting a diverse range of consumer needs and market segments. This includes developing affordable EVs, high-performance electric trucks, and fully autonomous commercial vehicles.

Enhancing Connected Vehicle Ecosystems

Ford is investing in connected vehicle ecosystems that integrate seamlessly with smart devices and digital lifestyles. This includes advancing FordPass capabilities, enhancing SYNC systems, and developing new connected services that offer personalized and intuitive user experiences.

Sustainability Leadership

Ford is committed to leading the industry in sustainability through comprehensive environmental initiatives, including carbon neutrality goals, sustainable manufacturing practices, and the adoption of circular economy principles. These efforts aim to minimize environmental impact and promote long-term sustainability.

Global Market Penetration

Ford plans to deepen its presence in emerging markets through localized production, strategic partnerships, and tailored product offerings. Expanding in regions with high growth potential ensures sustained revenue streams and enhanced global competitiveness.

Chevrolet's Future Strategic Initiatives

Diversification of Electric Vehicle Portfolio

Chevrolet is committed to diversifying its EV lineup with models that cater to various consumer preferences, including compact cars, SUVs, and trucks. This diversification ensures that Chevrolet can meet the diverse demands of the evolving electric mobility market.

Advancement of Autonomous Technologies

Chevrolet is accelerating the development and integration of autonomous driving technologies, enhancing Super Cruise capabilities, and exploring fully autonomous vehicle options. These advancements position Chevrolet as a leader in the autonomous driving space.

Commitment to Sustainability and Green Practices

Aligned with General Motors' sustainability goals, Chevrolet is focusing on reducing its carbon footprint, enhancing energy efficiency, and promoting the use of sustainable materials in vehicle production. These initiatives reinforce Chevrolet's commitment to environmental responsibility and sustainability.

Strengthening Global Presence

Chevrolet is expanding its global footprint by entering new markets, strengthening its presence in existing ones, and tailoring its product offerings to meet regional demands. Strategic collaborations and localized manufacturing are key components of this expansion strategy.

Collaborative Innovation and Industry Leadership

Both Ford and Chevrolet recognize the importance of collaborative innovation to stay ahead in a rapidly evolving market. Engaging in partnerships with technology firms, research institutions, and sustainability organizations enables both brands to leverage external expertise and accelerate their innovation efforts.

12. Anticipated Challenges and Mitigation Strategies

Technological Disruptions

The rapid pace of technological advancements poses challenges in terms of keeping up with innovation and integrating new technologies into existing platforms.

Mitigation Strategies:

- **Continuous R&D Investment:** Both Ford and Chevrolet are increasing investments in research and development to stay at the forefront of technological innovation.

- **Flexible Product Development:** Adopting flexible and agile product development processes allows for quicker integration of new technologies and features.

- **Talent Acquisition and Development:** Recruiting and nurturing talent with expertise in emerging technologies ensures that both brands have the necessary skills to drive innovation.

Supply Chain Resilience

Global supply chain disruptions, driven by geopolitical tensions, natural disasters, and pandemics, can impact production and delivery schedules.

Mitigation Strategies:

- **Localization of Production:** Establishing manufacturing facilities closer to key markets reduces dependency on international supply chains and enhances resilience.

- **Diversification of Suppliers:** Diversifying the

supplier base minimizes the risk of disruptions and ensures a steady flow of critical components.

- **Inventory Management:** Implementing advanced inventory management systems and strategies helps mitigate the impact of supply chain disruptions on production.

Regulatory Compliance and Environmental Standards

Meeting evolving regulatory requirements and environmental standards requires continuous adaptation and investment.

Mitigation Strategies:

- **Proactive Compliance Monitoring:** Establishing robust monitoring systems to stay abreast of regulatory changes and ensure timely compliance.

- **Sustainable Product Development:** Integrating sustainability into the product development process ensures that new models meet or exceed environmental standards.

- **Collaboration with Regulators:** Engaging in dialogue with regulatory bodies to influence policy development and stay informed about upcoming changes.

Consumer Acceptance and Adoption

The transition to electric and autonomous vehicles depends on consumer acceptance and willingness to adopt new technologies.

Mitigation Strategies:

- **Education and Awareness Campaigns:** Launching initiatives to educate consumers about the benefits and capabilities of electric and autonomous vehicles.

- **Incentives and Financing Options:** Offering incentives, rebates, and flexible financing options to make the transition to electric and autonomous vehicles more accessible.

- **Enhancing User Experience:** Ensuring that electric and autonomous vehicles offer superior performance, convenience, and safety features to encourage adoption.

Intense Competition from New Entrants

The entry of new competitors, particularly in the EV and autonomous driving sectors, intensifies the competitive landscape.

Mitigation Strategies:

- **Differentiation through Innovation:** Continuously innovating to offer unique features, superior performance, and enhanced user experiences that set Ford and Chevrolet apart from new entrants.

- **Strategic Partnerships:** Forming alliances with technology and sustainability leaders to enhance competitive capabilities and accelerate innovation.

- **Brand Loyalty Programs:** Strengthening brand loyalty through exceptional customer service, community engagement, and loyalty programs that reward repeat customers.

13. Conclusion: Navigating the Road Ahead

The future of the rivalry between Ford and Chevrolet is set to be shaped by a confluence of emerging technologies, market dynamics, and evolving consumer preferences. As both brands navigate the challenges and opportunities of the digital age, electrification, autonomy, sustainability, and global expansion will be pivotal in determining their trajectories.

Strategic Imperatives for Success

To thrive in the future automotive landscape, Ford and Chevrolet must:

- **Embrace Innovation:** Continuously invest in research and development to stay ahead of technological advancements and integrate cutting-edge features into their vehicles.

- **Prioritize Sustainability:** Commit to environmentally responsible practices and develop eco-friendly vehicles that meet the demands of a sustainability-conscious market.

- **Enhance Connectivity:** Develop advanced connected vehicle ecosystems that offer seamless integration with digital lifestyles and enhance user experiences.

- **Foster Community and Loyalty:** Engage with car enthusiasts, build strong communities, and cultivate brand loyalty through exceptional customer service and meaningful interactions.

- **Adapt to Market Shifts:** Remain agile and responsive to changing market conditions, consumer preferences, and regulatory requirements to maintain competitiveness and relevance.

Future Trajectory of the Rivalry

The rivalry between Ford and Chevrolet will continue to drive progress and innovation in the automotive industry. As they compete to lead in electrification, autonomous driving, and sustainable practices, both brands will push each other to achieve higher standards of performance, quality, and customer satisfaction. This competition benefits consumers by providing a diverse range of high-quality, innovative vehicles that cater to their evolving needs and preferences.

Legacy and Continued Influence

The enduring legacy of Ford and Chevrolet is built on their ability to adapt, innovate, and connect with consumers across generations. Their rivalry has not only defined their brands but also significantly influenced automotive trends and standards. As they look ahead, their continued commitment to excellence, sustainability, and technological leadership will ensure that their influence on the automotive industry remains profound and lasting.

Final Thoughts

The future of the Ford versus Chevrolet rivalry embodies the dynamic and transformative nature of the automotive industry. By anticipating and embracing emerging technologies, responding to market shifts, and maintaining a steadfast commitment to sustainability and innovation, Ford and Chevrolet will navigate the road ahead with resilience and vision. Their ongoing competition will drive the evolution of automotive excellence, shaping the future of mobility and reinforcing their positions as iconic leaders in the global automotive arena.

Conclusion: A Legacy Driven by Competition

The century-long rivalry between Ford and Chevrolet stands as one of the most enduring and influential competitions in the automotive industry. From their inception in the early 20th century to their present-day endeavors in electric and autonomous vehicles, Ford and Chevrolet have continually pushed each other to innovate, excel, and redefine the boundaries of automotive excellence. This conclusion reflects on how their competition has not only shaped the two companies but also left an indelible mark on the entire automotive landscape.

Summary: Shaping Companies and the Automotive Industry

Catalysts for Innovation and Excellence

The relentless competition between Ford and Chevrolet has been a primary catalyst for innovation within both companies. Each new model, from the iconic Ford Mustang and Chevrolet Camaro to the robust Ford F-150 and Chevrolet Silverado, has been a testament to their commitment to performance, design, and technological advancement. This rivalry has spurred continuous improvements in vehicle safety, fuel efficiency, and overall performance, ensuring that both brands remain at the forefront of automotive engineering.

Driving Market Expansion and Global Presence

Ford and Chevrolet's competition has extended beyond domestic markets, driving both companies to pursue aggressive global expansion strategies. Their efforts to penetrate high-potential regions in Asia, Europe, and South America have not only increased their market share but also introduced diverse consumer preferences and regulatory challenges. This global push has fostered a deeper understanding of international markets, leading to more adaptable and resilient business models that have benefited the broader automotive industry.

Influence on Marketing and Community Building

The rivalry has significantly influenced marketing strategies and community engagement practices. Both brands have harnessed the power of digital marketing, social media, and fan communities to build strong, loyal customer bases. Iconic marketing campaigns and strategic use of pop culture—such as the Ford Mustang's role in "Bullitt" and the Chevrolet Camaro's portrayal as Bumblebee in "Transformers"—have cemented their status as cultural icons. These efforts have not only enhanced brand visibility but also fostered a sense of belonging and pride among enthusiasts, driving sustained brand loyalty.

Advancements in Sustainability and Technology

As the automotive industry transitions towards sustainability and advanced technologies, Ford and Chevrolet have been at the forefront of this evolution. Their investments in electric vehicles (EVs), autonomous driving technologies, and sustainable manufacturing practices have set benchmarks for the industry. By embracing these advancements, both companies have demonstrated their commitment to environmental responsibility and technological leadership, influencing industry standards and consumer expectations.

Economic Impact and Aftermarket Growth

The competition between Ford and Chevrolet has also had a profound economic impact, fueling the growth of the aftermarket industry. Car enthusiasts, inspired by these iconic brands, have invested heavily in customizing and personalizing their vehicles, creating a thriving ecosystem of performance parts, aesthetic modifications, and specialized services. This aftermarket growth has not only generated significant revenue but also driven innovation and diversity within the automotive sector.

Resilience Through Challenges and Controversies

Despite facing numerous recalls, legal battles, and public controversies, both Ford and Chevrolet have demonstrated remarkable resilience. Their ability to manage crises effectively, prioritize customer safety and satisfaction, and implement systemic improvements has reinforced their reputations and ensured long-term sustainability. These experiences have underscored the importance of transparency, accountability, and continuous improvement in maintaining brand integrity and consumer trust.

Final Thoughts: The Enduring Power of Competition

The rivalry between Ford and Chevrolet exemplifies the enduring power of competition in driving progress and fostering excellence. This competition has not only shaped the destinies of these two automotive giants but has also been instrumental in shaping the broader automotive industry. The lessons learned from their century-long battle highlight several key principles that are essential for sustained success in any competitive landscape:

1. Continuous Innovation

Competition compels companies to innovate relentlessly. Ford and Chevrolet's drive to outperform each other has led to significant advancements in vehicle design, performance, and technology, setting higher standards for the entire industry.

2. Customer-Centric Approach

Understanding and responding to customer needs and

preferences are crucial. Both brands have thrived by prioritizing customer satisfaction, offering vehicles that resonate with diverse consumer segments, and fostering strong emotional connections through community engagement and cultural integration.

3. Adaptability and Resilience

The ability to adapt to changing market conditions, regulatory landscapes, and technological advancements is vital. Ford and Chevrolet have demonstrated resilience by evolving their strategies, embracing new technologies, and expanding their global footprints to stay relevant and competitive.

4. Strategic Partnerships and Collaboration

Collaborative efforts with technology firms, sustainability organizations, and global partners have been pivotal in driving innovation and enhancing operational efficiency. These partnerships enable both companies to leverage external expertise and accelerate their progress in key areas.

5. Commitment to Sustainability

As environmental concerns gain prominence, a steadfast commitment to sustainability is imperative. Ford and Chevrolet's investments in electric vehicles, sustainable manufacturing practices, and carbon neutrality goals reflect their dedication to responsible stewardship and future-proofing their businesses.

6. Leveraging Pop Culture and Community

Engaging with pop culture and building vibrant communities around their brands have strengthened Ford and Chevrolet's market presence and fostered deep brand loyalty. These efforts highlight the importance of cultural relevance and community in driving brand success.

7. Effective Crisis Management

Navigating recalls, legal challenges, and public controversies with transparency and accountability has reinforced trust and demonstrated the importance of robust crisis management frameworks in sustaining long-term brand integrity.

8. Economic and Aftermarket Synergy

The symbiotic relationship between primary vehicle sales and the aftermarket industry underscores the broader economic impact of automotive competition. Supporting enthusiast communities and aftermarket growth not only drives additional revenue but also enhances brand engagement and loyalty.

Looking Forward: The Path to Continued Excellence

As we look to the future, the rivalry between Ford and Chevrolet will undoubtedly continue to shape the automotive landscape. Emerging technologies such as artificial intelligence, blockchain, augmented reality, and advanced manufacturing processes will further redefine vehicle capabilities and consumer experiences. The ongoing shift towards electrification and autonomy will present both challenges and opportunities, pushing Ford and Chevrolet to innovate and adapt in ways that will benefit consumers and the industry alike.

Moreover, the increasing importance of sustainability and corporate social responsibility will drive both brands to pioneer eco-friendly practices and technologies, contributing to a greener and more sustainable automotive future. Their ability to balance heritage with modernity, embrace innovation while maintaining brand identity, and foster strong community connections will be key to their continued success.

In essence, the century-long rivalry between Ford and Chevrolet exemplifies how competition can be a powerful force for good—driving innovation, enhancing quality, and expanding consumer choices. This enduring competition not only fuels the growth and evolution of these two iconic brands but also propels the entire automotive industry towards greater heights of excellence and sustainability.

As Ford and Chevrolet navigate the road ahead, their commitment to innovation, customer satisfaction, and sustainability will ensure that their legacy remains strong and influential. The story of their rivalry is a testament to the transformative power of competition, illustrating how it can shape not only the destinies of individual companies but also the broader trajectory of an entire industry.

About the Author

Etienne Psaila, an accomplished author with over two decades of experience, has mastered the art of weaving words across various genres. His journey in the literary world has been marked by a diverse array of publications, demonstrating not only his versatility but also his deep understanding of different thematic landscapes. However, it's in the realm of automotive literature that Etienne truly combines his passions, seamlessly blending his enthusiasm for cars with his innate storytelling abilities.

Specializing in automotive and motorcycle books, Etienne brings to life the world of automobiles through his eloquent prose and an array of stunning, high-quality color photographs. His works are a tribute to the industry, capturing its evolution, technological advancements, and the sheer beauty of vehicles in a manner that is both informative and visually captivating.

A proud alumnus of the University of Malta, Etienne's academic background lays a solid foundation for his meticulous research and factual accuracy. His education has not only enriched his writing but has also fueled his career as a dedicated teacher. In the classroom, just as in his writing, Etienne strives to inspire, inform, and ignite a passion for learning.

As a teacher, Etienne harnesses his experience in writing to engage and educate, bringing the same level of dedication and excellence to his students as he does to his readers. His dual role as an educator and author makes him uniquely positioned to understand and convey complex concepts with clarity and ease, whether in the classroom or through the pages of his books.

Through his literary works, Etienne Psaila continues to leave an indelible mark on the world of automotive literature, captivating car enthusiasts and readers alike with his insightful perspectives and compelling narratives.

Visit www.etiennepsaila.com for more.